A Mathematical Theory of Arguments for Statistical Evidence

Contributions to Statistics

V. Fedorov/W.G. Müller/I.N. Vuchkov (Eds.)
Model-Oriented Data Analysis,
XII/248 pages, 1992

J. Antoch (Ed.)
Computational Aspects of Model Choice,
VII/285 pages, 1993

W.G. Müller/H.P. Wynn/A.A. Zhigljavsky (Eds.)
Model-Oriented Data Analysis,
XIII/287 pages, 1993

P. Mandl/M. Hušková (Eds.)
Asymptotic Statistics,
X/474 pages, 1994

P. Dirschedl/R. Ostermann (Eds.)
Computational Statistics,
VII/553 pages, 1994

C.P. Kitsos/W.G. Müller (Eds.)
MODA 4 – Advances in Model-Oriented Data Analysis,
XIV/297 pages, 1995

H. Schmidli
Reduced Rank Regression,
X/179 pages, 1995

W. Härdle/M.G. Schimek (Eds.)
Statistical Theory and Computational Aspects of Smoothing,
VIII/265 pages, 1996

S. Klinke
Data Structures for Computational Statistics,
VIII/284 pages, 1997

A.C. Atkinson/L. Pronzato/H.P. Wynn (Eds.)
MODA 5 – Advances in Model-Oriented Data Analysis and Experimental Design,
XIV/300 pages, 1998

M. Moryson
Testing for Random Walk Coefficients in Regression and State Space Models,
XV/317 pages, 1998

S. Biffignandi (Ed.)
Micro- and Macrodata of Firms,
XII/776 pages, 1999

W. Härdle/Hua Liang/J. Gao
Partially Linear Models,
X/203 pages, 2000

Paul-André Monney

A Mathematical Theory of Arguments for Statistical Evidence

Springer-Verlag Berlin Heidelberg GmbH

Series Editors
Werner A. Müller
Martina Bihn

Author
Paul-André Monney
Purdue University
Department of Statistics
1399 Mathematical Sciences Building
West Lafayette, IN 47907-1399
USA
monney@stat.purdue.edu

ISSN 1431-1968
ISBN 978-3-7908-1527-6 ISBN 978-3-642-51746-4 (eBook)
DOI 10.1007/978-3-642-51746-4

Cataloging-in-Publication Data applied for
Die Deutsche Bibliothek – CIP-Einheitsaufnahme
Monney, Paul-André: A mathematical theory of arguments for statistical evidence / Paul-André Monney. – Heidelberg: Physica-Verl., 2003
(Contributions to statistics)

Softcover Design: Erich Kirchner, Heidelberg

SPIN 10890114 88/2202-5 4 3 2 1 0 – Printed on acid-free and non-aging paper

To Brenda

Preface

The subject of this book is the reasoning under uncertainty based on statistical evidence, where the word reasoning is taken to mean searching for arguments in favor or against particular hypotheses of interest. The kind of reasoning we are using is composed of two aspects. The first one is inspired from classical reasoning in formal logic, where deductions are made from a knowledge base of observed facts and formulas representing the domain specific knowledge. In this book, the facts are the statistical observations and the general knowledge is represented by an instance of a special kind of statistical models called functional models. The second aspect deals with the uncertainty under which the formal reasoning takes place. For this aspect, the theory of hints [27] is the appropriate tool. Basically, we assume that some uncertain perturbation takes a specific value and then logically evaluate the consequences of this assumption. The original uncertainty about the perturbation is then transferred to the consequences of the assumption. This kind of reasoning is called assumption-based reasoning. Before going into more details about the content of this book, it might be interesting to look briefly at the roots and origins of assumption-based reasoning in the statistical context.

In 1930, R.A. Fisher [17] defined the notion of fiducial distribution as the result of a new form of argument, as opposed to the result of the older Bayesian argument. The idea of functional model was implicitly present in the fiducial argument, as shown by the analysis of Dawid and Stone [10]. Several other authors, like e.g. Bunke [7] and Plante [35], have considered the fiducial argument from a functional perspective. On the other hand, Dempster [14] presented a theory that was broad enough to include both the fiducial argument and the standard Bayesian methods. A key concept introduced by Dempster was the notion of upper and lower probabilities induced by a multivalued mapping [12]. Some of these ideas were later used by Shafer [38] and resulted in a mathematical theory of evidence using belief and plausibility functions. Furthermore, Kohlas & Monney [27] reinterpreted the Dempster-Shafer theory of evidence which resulted in the theory of hints.

In view of this historical development, it is not surprising that the theory of hints is the appropriate tool to represent the result of assumption-based reasoning on functional models and their generalization. Each hint about the unknown value of the parameter permits to evaluate some hypothesis of interest by computing its degree of support and its degree of plausibility. The degree of support measures the strength of the arguments speaking in favor of the hypothesis, whereas the degree of plausibility measures the degree of compatibility between the hypothesis and the available evidence. The plausibility function gives a clear and natural interpretation to the likelihood function. Along with the concept of hint, the notion of generalized functional model and its analysis by assumption-based reasoning is central to this book.

In the first two chapters, the theory of generalized functional models for a discrete parameter is developed. This theory is used to define a generalized notion of weight of evidence. Obviously, such a concept has many possible fields of applications, ranging from law (e.g. in investigating the statistical evidence of discrimination), to business (e.g. in evaluating various competing economic hypotheses) and medicine (e.g. in evaluating the merits of treatments based on clinical trials).

The second part of the book is dedicated to the study of special linear functional models called Gaussian linear models. Since these models involve continuous spaces, we need to develop a theory of continuous hints, which is done in chapter 3. In the following chapters, several kinds of Gaussian linear systems are analyzed, each time resulting in a Gaussian hint for the parameter. In particular, the analysis of the classical linear regression model is performed.

As there are usually several different sources of information for some related questions, the problem of how to pool several Gaussian hints is addressed. Then, by properly defining the concept of a marginal hint, it is shown that Gaussian hints form a valuation system in the sense of Shenoy and Shafer [44] for which the so-called local propagation axioms are satisfied. This allows for an efficient computation of the inferred hint about the parameter vector of interest. It should be mentioned that valuation systems and their axioms constitute one of the most important achievements in uncertain reasoning and Gaussian hints form a further instance of this general framework. The local propagation algorithm is presented in chapter 7.

Finally, in chapter 8, an application of our theory to a problem in control theory is considered. More precisely, it is shown that the celebrated Kalman filter [23] can be derived from our approach by locally propagating Gaussian hints in a Markov tree.

It is my pleasure to acknowledge a number of people who have helped to bring this book into existence. I am deeply indebted to Prof. J. Kohlas for his many suggestions and ideas which are at the origin of this work. I also want to thank him for his constant support and interest in my research.

I also wish to express my gratitude to Prof. H.W. Brachinger and Prof. A.P. Dempster for their support and encouragement. I am thankful to Prof. P.P. Shenoy and Prof. P. Smets for the many interesting and stimulating discussions we had during the time they spent at the University of Fribourg.

My present and past friends and colleagues of the Seminar of Statistics, the Department of Quantitative Economics and the Department of Informatics of the University of Fribourg also deserve my gratitude for their help and for the friendly working atmosphere I have enjoyed during so many years. I also want to thank the Swiss Federal Office for Education and Science for their financial support.

Most of all, my gratitude goes to my wife Brenda, who provided me with her constant support, enthusiasm and encouragement during the preparation of this book.

Fribourg, July 2002 *Paul-André Monney*

Contents

1. The Theory of Generalized Functional Models

In this chapter we present the basics of a new theory called the theory of generalized functional models. By discussing several examples, this theory is shown to be very natural for modeling various situations of reasoning under uncertainty. One of its major advantages is that it enables us to better represent the information that is really available at the beginning of the analysis. For example, there is no need to artificially create prior information on the variables of interest when such information does not exist. But if some information is available, then it can easily be included in the model.

Definition 1 *A generalized functional model is a pair* (f, P) *where* f *is a function describing the interactions between a parameter variable, an observation variable and a random source, whereas* P *is a probability distribution for the random source.*

Unlike traditional functional models, generalized functional models do not require that only one value of the parameter variable is compatible with an observation and a realization of the random source. As a consequence, the results of the analysis of a generalized functional model are not expressed in terms of probability distributions but rather by mathematical structures called hints and their associated support and plausibility functions. The theory of generalized functional models is very logical and is inspired by ideas put forward by R.A. Fisher in his theory of fiducial probability [17].

As the theory of hints plays a central role in the analysis of a generalized functional model, as in the theory of Gaussian linear systems which will later be discussed, it is quite natural that the first section of this chapter provides a brief introduction to this theory by presenting its most prominent concepts and properties.

1.1 The Theory of Hints on Finite Frames

The theory of hints developed by Kohlas and Monney [27, 26] is a variant of the Dempster-Shafer theory of evidence [12, 38]. As demonstrated later on, this theory plays an important role in this book.

1.1.1 Definition of a Hint and its Associated Functions

Suppose that a question is under investigation and its unique answer θ^* is unknown. If Θ denotes the set of all possible answers to the question, then we obviously have $\theta^* \in \Theta$. Now suppose that Θ, the so-called frame of discernment, is linked with another set Ω through a known multivalued mapping

$$\Gamma : \Omega \longrightarrow \mathcal{P}(\Theta)$$

where $\mathcal{P}(\Theta)$ denotes the power set of Θ. Also, there is exactly one unknown correct value in Ω and it is assumed that we are in a position to specify a probability measure P on Ω with $P(\omega)$ describing the chance that ω in Ω is the correct value. The elements of Ω are sometimes called *assumptions*.

The mapping Γ is constructed from some piece of information we have about the question under investigation. More precisely, under the assumption that ω in Ω is correct, that piece of information permits to conclude that the correct answer to the question, namely θ^*, will inevitably be an element of $\Gamma(\omega)$. This set is chosen to be as small as possible, but, depending on the situation, it may very well happen that it is the entire frame Θ itself. This case occurs when the assumption ω does not permit to restrict the set of possible answers to the question. By construction, the so-called *focal sets* $\Gamma(\omega)$, $\omega \in \Omega$ are not empty. We assume for now that both Θ and Ω are finite sets. Note that this kind of reasoning by assumption is also considered by Laskey & Lehner for models in artificial intelligence [28].

Definition 2 *A structure of the form*

$$\mathcal{H} = (\Omega, P, \Gamma, \Theta)$$

is called a hint for Θ.

In the light of a piece of information about Θ represented by a hint, we are interested in the evaluation of the hypothesis that θ^* is in a particular subset H of Θ. In other words, what does the hint say about the hypothesis H, e.g. is it highly plausible ?

Under the assumption that ω is correct, it is certain that $\theta^* \in \Gamma(\omega)$. So, if $\Gamma(\omega) \subseteq H$, then θ^* is in H and so H is true. Hence, each ω such that $\Gamma(\omega) \subseteq H$ speaks in favor, supports the hypothesis H. Let $u(H)$ denote this set of supporting assumptions, i.e.

$$u(H) = \{\omega \in \Omega : \Gamma(\omega) \subseteq H\}.$$

Since $P(u(H))$ is the probability that the correct assumption supports H, it is natural to define the *degree of support* of H by

$$sp(H) = P(u(H)).$$

We may also ask to what extent does the hypothesis H appear as possible or plausible in view of the hint. Since the hypothesis H remains possible under an assumption ω if and only if $\Gamma(\omega) \cap H$ is not empty, it is natural to define the *degree of plausibility* of H by

$$pl(H) = P(v(H))$$

where

$$v(H) = \{\omega \in \Omega : \Gamma(\omega) \cap H \neq \emptyset\}. \tag{1.1.1}$$

The value of $pl(H)$ expresses the degree of compatibility between the hypothesis H and the available knowledge represented by the hint. The corresponding functions

$$sp : \mathcal{P}(\Theta) \longrightarrow [0,1]$$

and

$$pl : \mathcal{P}(\Theta) \longrightarrow [0,1]$$

are respectively called *support function* and *plausibility function.*

Theorem 1 *Let H, H' and H'' be subsets of Θ. Then*

1. $sp(\emptyset) = pl(\emptyset) = 0, \quad sp(\Theta) = pl(\Theta) = 1$
2. $sp(H) = 1 - pl(H^c), \quad pl(H) = 1 - sp(H^c)$
3. $sp(H) \leq pl(H)$
4. *If $H' \subseteq H''$ then $sp(H') \leq sp(H'')$ and $pl(H') \leq pl(H'')$*
5. *Let $H_i \subseteq \Theta$, $i = 1, \ldots, n$ be a collection of subsets of Θ. If we define*

$$S = \{I \subseteq \{1, \ldots, n\} : I \neq \emptyset\},$$

then

$$sp(\cup_{i=1}^{n} H_i) \geq \sum \{(-1)^{|I|+1} sp(\cap_{i \in I} H_i) : I \in S\}.$$

Proof. This theorem is proved in Kohlas & Monney [27]. □

The second part of property 2 expresses the idea that a hypothesis appears as less plausible as there is more support for its negation. Property 5 states that support functions are not necessarily additive measures, i.e. we can have

$$sp(H) + sp(H^c) < 1.$$

Definition 3 *A hint is called precise if all of its focal sets are singletons.*

Theorem 2 *Let sp and pl denote the support and plausibility functions of a precise hint. Then*

1. *$sp(H) = pl(H)$ for all $H \subseteq \Theta$, i.e. sp and pl coincide.*
2. *sp is a probability measure on $\mathcal{P}(\Theta)$.*

Proof. This theorem is proved in Kohlas & Monney [27]. □

Definition 4 *Given a hint $\mathcal{H}$, the mapping*

$$m : \mathcal{P}(\Theta) \longrightarrow [0,1]$$

given by

$$m(A) = P(\{\omega \in \Omega : \Gamma(\omega) = A\})$$

is called the basic probability assignment, or m-function, of $\mathcal{H}$.

Note that $m(A) = 0$ if A is not a focal set because in this case there is no ω such that $\Gamma(\omega) = A$.

It can easily be proved that

$$m(\emptyset) = 0,$$

$$\sum \{m(A) : A \subseteq \Theta\} = 1,$$

$$sp(H) = \sum \{m(A) : A \subseteq H\},$$

and

$$pl(H) = \sum \{m(A) : A \cap H \neq \emptyset\}.$$

This shows that the m-function can be used to compute degrees of support and degrees of plausibility and that a support function is a belief function in the sense of Shafer [38]. In addition, if the hint is precise, we have

$$m(\theta) = sp(\theta) = pl(\theta)$$

for all $\theta \in \Theta$, which implies that m is a probability distribution on Θ.

Definition 5 *For precise hints, the restriction of m to the singletons is called the probability density of the hint.*

The motivation for using the term probability density comes from the continuous case to be considered in chapter 3. Two hints having the same m-function are called *equivalent* because they generate the same support and plausibility functions.

If we define

$$w(H) = \{\omega \in \Omega : H \subseteq \Gamma(\omega)\},$$

then

$$q(H) = P(w(H))$$

is called the *commonality* of H and the corresponding mapping

$$q : \mathcal{P}(\Theta) \longrightarrow [0,1]$$

is called the *commonality function* of the hint. The notation q for the commonality function is taken from Shafer [38]. Of course, we have

$$q(H) = \sum \{m(A) : H \subseteq A\} \tag{1.1.2}$$

and it can be proved that $q(\theta) = pl(\theta)$ for all $\theta \in \Theta$ [27].

1.1.2 Combination of Hints

If there are several hints relative to the same frame of discernment, they need to be combined to get a new hint representing the combined information. Let

$$\mathcal{H}_i = (\Omega_i, P_i, \Gamma_i, \Theta), \quad i = 1, \ldots, n$$

be a family of hints. The finite set

$$\Omega_1 \times \ldots \times \Omega_n$$

is the set of all combined assumptions.

Definition 6 *The hints $\mathcal{H}_i$ are called independent if, for all combined assumptions $\omega = (\omega_1, \ldots, \omega_n)$, the probability that ω is correct is*

$$\prod_{i=1}^{n} P_i(\omega_i).$$

In other words, the hints $\mathcal{H}_i$ are independent if their associated random assumptions $\tilde{\omega}_i$ are independent random variables. From now on we assume that the hints we combine are independent. Clearly, because of the meaning of the multivalued mappings Γ_i, a combined assumption $(\omega_1, \ldots, \omega_n)$ such that

$$\cap_{i=1}^{n} \Gamma_i(\omega_i) = \emptyset$$

is impossible and

$$k = \sum \{\prod_{i=1}^{n} P_i(\omega_i) : \cap_{i=1}^{n} \Gamma_i(\omega_i) = \emptyset\} \tag{1.1.3}$$

is the probability of getting an impossible combined assumption. Since such assumptions cannot be correct, we have to condition the product measure

$$\prod_{i=1}^{n} P_i$$

on the event

$$\Omega = \{(\omega_1, \ldots, \omega_n) \in \Omega_1 \times \ldots \times \Omega_n : \cap_{i=1}^{n} \Gamma_i(\omega_i) \neq \emptyset\},$$

provided that the probability of Ω is not zero, i.e. $k < 1$. In this case, if P denotes the new probability measure on Ω, then we have

$$P(\omega) = \frac{\prod_{i=1}^{n} P_i(\omega_i)}{1 - k}$$

for all $\omega = (\omega_1, \ldots, \omega_n) \in \Omega$. For $\omega \in \Omega$, let

$$\Gamma(\omega) = \cap_{i=1}^{n} \Gamma_i(\omega_i)$$

denote the non-empty focal set associated with ω. In other words, if ω is the correct combined assumption, then θ^* is inevitably in $\Gamma(\omega)$.

Definition 7 *If $k < 1$, then the hints $\mathcal{H}_1, \ldots, \mathcal{H}_n$ can be combined and the hint $\mathcal{H} = (\Omega, P, \Gamma, \Theta)$ is called the combination of $\mathcal{H}_1, \ldots, \mathcal{H}_n$ and we write*

$$\mathcal{H} = \mathcal{H}_1 \oplus \ldots \oplus \mathcal{H}_n.$$

This procedure to combine hints is called Dempster's rule of combination because it was introduced by A.P. Dempster in 1967 [12].

Now we present several useful results on the combination of hints.

Theorem 3 *Let $\mathcal{H}_1, \ldots, \mathcal{H}_n$ be a family of hints relative to the frame Θ and let m_i denote the m-function of $\mathcal{H}_i$. Then the value of k defined in equation (1.1.3) is*

$$k = \sum \{\prod_{i=1}^{n} m_i(A_i) : A_i \subseteq \Theta, \cap_{i=1}^{n} A_i = \emptyset\}$$

and the combined hint

$$\mathcal{H} = \mathcal{H}_1 \oplus \ldots \oplus \mathcal{H}_n$$

exists if $k < 1$. In this case, the m-function of $\mathcal{H}$ is

$$m(A) = \frac{\sum \{\prod_{i=1}^{n} m_i(A_i) : A_i \subseteq \Theta, \cap_{i=1}^{n} A_i = A\}}{1-k} \tag{1.1.4}$$

for all non-empty subsets $A \subseteq \Theta$ and of course $m(\emptyset) = 0$.

Proof. The proof of this theorem can be found in Kohlas & Monney [27]. □

Up to equivalence, the combination of hints is both associative and commutative [27].

Theorem 4 *Let $\mathcal{H}_1, \ldots, \mathcal{H}_n$ be a collection of hints with commonality functions $q_1, \ldots, q_n$ respectively. If q denotes the commonality function of the combined hint*

$$\mathcal{H} = \mathcal{H}_1 \oplus \ldots \oplus \mathcal{H}_n,$$

then

$$q(A) = \frac{\prod_{i=1}^{n} q_i(A)}{1-k}$$

for all non-empty subset $A \subseteq \Theta$.

Proof. The proof of this result can be found in Kohlas & Monney [27]. □

Theorem 5 *Let $\mathcal{H}_0$ be a precise hint on Θ and $\mathcal{H}_1$ an arbitrary hint on Θ. If m_0 is the m-function of $\mathcal{H}_0$ and pl_1 is the plausibility function of $\mathcal{H}_1$, then $\mathcal{H}_0 \oplus \mathcal{H}_1$ is a precise hint with probability density*

$$m(\theta) = c\, m_0(\theta) pl_1(\theta)$$

where

$$c = \left(\sum_{\theta \in \Theta} m_0(\theta) pl_1(\theta) \right)^{-1}.$$

Proof. The proof of this result can be found in Kohlas & Monney [27]. □

Theorem 6 *If $\mathcal{H}_1, \ldots, \mathcal{H}_n$ are precise hints on Θ with probability densities $m_1, \ldots m_n$, then $\oplus_{i=1}^n \mathcal{H}_i$ is again a precise hint with probability density*

$$m(\theta) = c \prod_{i=1}^{n} m_i(\theta)$$

where

$$c = \left(\sum_{\theta \in \Theta} \prod_{i=1}^{n} m_i(\theta) \right)^{-1}.$$

Proof. The proof of this result can be found in Kohlas & Monney [27]. □

Theorem 7 *Let m_0 be the probability density of a precise hint $\mathcal{H}$. If we define $\mathcal{H}_i = \mathcal{H}$ for $i = 1, \ldots, n$, then the probability density of the precise combined hint $\oplus_{i=1}^n \mathcal{H}_i$ is given by*

$$m(\theta) = \frac{m_0(\theta)^n}{\sum_{\theta \in \Theta} m_0(\theta)^n}$$

for all $\theta \in \Theta$.

Proof. This result is a direct consequence of theorem 6. □

1.1.3 Mass functions and gem-functions

We have seen that an m-function is a mapping

$$m : \mathcal{P}(\Theta) \longrightarrow [0, 1]$$

satisfying the two conditions

$$m(\emptyset) = 0 \tag{1.1.5}$$

$$\sum \{m(A) : A \subseteq \Theta\} = 1. \tag{1.1.6}$$

An m-function is a function that is derived from a hint. However, if we define a *mass function* as a mapping

$$m : \mathcal{P}(\Theta) \longrightarrow [0, 1]$$

satisfying equations (1.1.5) and (1.1.6), then Shafer [38] defines the *degree of belief* of a hypothesis $H \subseteq \Theta$ as

$$Bel(H) = \sum \{m(A) : A \subseteq H\},$$

and the corresponding mapping

$$Bel : \mathcal{P}(\Theta) \longrightarrow [0,1]$$

is called a *belief function* by Shafer [38]. Of course, every m-function is a mass function. The notion of mass function has been generalized by Smets by allowing the possibility to assign a positive mass to the empty set, which leads to the concepts of basic belief assignment and non-normalized belief function [48]. In this book, such a generalized mass function will be called a gem-function :

Definition 8 *A gem-function g on a frame Θ is a mapping*

$$g : \mathcal{P}(\Theta) \longrightarrow [0,1] \tag{1.1.7}$$

such that

$$\sum \{g(A) : A \subseteq \Theta\} = 1. \tag{1.1.8}$$

The difference with a mass function is that a gem-function may assign a positive mass to the empty set, which is not possible for a mass function. Of course, every mass function, in particular every m-function, is a gem-function but the converse is false. Also, a gem-function is called *proper* if the value assigned to the empty set is strictly smaller than 1. Note that a proper gem-function g can be transformed into an mass function m by normalization :

$$m(A) = \begin{cases} \frac{g(A)}{1-g(\emptyset)} & \text{if } \emptyset \neq A \subseteq \Theta \\ 0 & \text{if } A = \emptyset. \end{cases}$$

By equation (1.1.2), the commonality function of a hint or of its m-function m is the mapping

$$q : \mathcal{P}(\Theta) \longrightarrow [0,1] \tag{1.1.9}$$

defined by

$$q(H) = \sum \{m(A) : H \subseteq A\}. \tag{1.1.10}$$

Similarly, for a gem-function g, the commonality function of g is the mapping

$$q : \mathcal{P}(\Theta) \longrightarrow [0,1] \tag{1.1.11}$$

defined by

$$q(H) = \sum \{g(A) : H \subseteq A\}. \tag{1.1.12}$$

Obviously, the definitions given in equations (1.1.10) and (1.1.12) coincide if the gem-function g happens to be an m-function. Note that the commonality function is also called the q-function

Smets [47] considers the combination of several gem-functions defined on a frame Θ. Since the present book is placed in the context of the Theory of Hints and more generally the Dempster-Shafer Theory of Evidence, the combination of gem-functions is considered here only as a technical tool, whereas for Smets it is an essential component of his Transferable Belief

Model with its own meaning and interpretation. For this reason, we do not speak of the combination of several gem-functions, but we rather talk about the gem-function associated with a collection of gem-functions.

Definition 9 *Let*

$$C = \{g_1, \ldots, g_n\}$$

be a collection of gem-functions on Θ that are not necessarily different, i.e. some gem-functions may appear several times in the collection. The gem-function associated with the collection C is the mapping

$$g : \mathcal{P}(\Theta) \longrightarrow [0,1]$$

given by

$$g(A) = \sum \{\prod_{i=1}^{n} g_i(A_i) : A_i \subseteq \Theta, \cap_{i=1}^{n} A_i = A\}$$

for all subsets A of Θ. The gem-function g associated with C is denoted by $gem(C)$ and to simplify the notation we simply write $gem(g_1, \ldots, g_n)$ instead of $gem(\{g_1, \ldots, g_n\})$.

The mapping $gem(C)$ clearly satisfies conditions (1.1.7) and (1.1.8) and does not depend on the order in which the elements of C are considered. As the following definition shows, this allows us to speak about the gem-function associated with a collection of hints because every m-function is a mass function and hence also a gem-function.

Definition 10 *For $i = 1, \ldots, n$, let m_i denote the m-function of the hint $\mathcal{H}_i$ on Θ. Then the gem-function associated with the collection of hints*

$$C = \{\mathcal{H}_1, \ldots, \mathcal{H}_n\}$$

is

$$gem(C) = gem(m_1, \ldots, m_n).$$

By theorem 3, the combined hint

$$\mathcal{H} = \mathcal{H}_1 \oplus \ldots \oplus \mathcal{H}_n$$

exists if

$$gem(C)(\emptyset) < 1,$$

in which case its m-function is obtained by normalization of the gem-function $gem(C)$. Therefore, to find the m-function of the combination of several hints, we need to find a method for computing $gem(g_1, \ldots, g_n)$. For this purpose, the notion of Dempster specialization matrix is considered in the next section.

1.1.4 Dempster Specialization Matrices

From now on, let

$$\{B_1, B_2, \ldots, B_n\}$$

denote the set of all subsets of Θ and it is assumed that the following conditions are satisfied :

$$B_1 = \emptyset \quad \text{and} \quad (B_i \subseteq B_j \ \Rightarrow\ i \leq j). \tag{1.1.13}$$

Note that it is always possible to find such an ordering of the subsets of Θ. A gem-function g is then completely specified by a column vector

$$g = (g(B_1), \ldots, g(B_n))'$$

and, similarly, a commonality function q is completely specified by a column vector

$$q = (q(B_1), \ldots, q(B_n))',$$

where the prime denotes the transpose of the vector. The following notion of Dempster specialization matrix introduced by Klawonn and Smets [24] will be useful in the sequel.

Definition 11 *Let g be a gem-function on Θ. Then the Dempster specialization matrix of g is the square matrix S of order n given by*

$$S_{ij} = \sum \{g(K) : K \subseteq \Theta, K \cap B_j = B_i\} \tag{1.1.14}$$

for all i and j in $\{1, \ldots, n\}$.

If $U = B_i$ and $V = B_j$, then the element S_{ij} of the matrix S is also denoted by

$$S(U, V) = \sum \{g(K) : K \cap V = U\}.$$

Also, in order to keep the notation as simple as possible, it will be sometimes useful to simply write i instead of B_i and j instead of B_j. With this convention, the equation

$$S_{ij} = \sum \{g(k) : k \subseteq \Theta, k \cap j = i\}$$

still makes sense.

Basically, a special case of the following result is first mentioned in Klawonn and Smets [24], but unfortunately no proof is given there.

Theorem 8 *Let $\{g_1, \ldots, g_n, g_{n+1}\}$ be a collection of gem-functions on Θ. If S denotes the Dempster specialization matrix of g_{n+1}, then*

$$gem(g_1, \ldots, g_{n+1}) = S \cdot gem(g_1, \ldots, g_n).$$

Proof. The proof of this result can be found in [32]. □

1.1.5 A Representation of the Dempster Specialization Matrix

In this section it will be shown that the Dempster specialization matrix can be diagonalized.

Definition 12 *The incidence matrix of the ordering $B_1, \dots, B_n$ is the square matrix M of order n given by*

$$M_{ij} = \begin{cases} 1 & \text{if } B_i \subseteq B_j \\ 0 & \text{otherwise} \end{cases} \tag{1.1.15}$$

for all i and j in $\{1, \dots, n\}$.

The matrix M is an upper-triangular matrix because if $i > j$ then $M_{ij} = 0$ because $B_i \not\subseteq B_j$ by the second condition in (1.1.13). Also, M is a regular matrix because $M_{ii} = 1$ for all $i = 1, \dots, n$. Now let g be a gem-function given by the column vector

$$g = (g(B_1), \dots, g(B_n))'.$$

If the commonality function of g is given by the column vector

$$q = (q(B_1), \dots, q(B_n))',$$

then we obviously have

$$Mg = q. \tag{1.1.16}$$

Definition 13 *Let q denote the commonality function of a gem-function g on the frame Θ. Then the commonality matrix of g is the square matrix Q of order n given by*

$$Q_{ij} = \begin{cases} q(B_i) & \text{if } i = j \\ 0 & \text{otherwise} \end{cases} \tag{1.1.17}$$

for all i and j in $\{1, \dots, n\}$.

The following theorem states that the Dempster specialization matrix of a gem-function is diagonalizable. This result is first mentioned in Klawonn and Smets [24], but unfortunately no proof is given there.

Theorem 9 *Let S denote the Dempster specialization matrix and Q the commonality matrix of a gem-function g on the frame Θ. If M is the incidence matrix, then*

$$S = M^{-1}QM.$$

Proof. The proof of this result can be found in [32]. □

When a matrix is diagonalizable, then a certain number of general statements can be proved (see e.g. Harville [21]). When they are applied to a Dempster specialization matrix, these results are expressed as follows.

If $r_1, \dots, r_n$ are the n column vectors of the incidence matrix M^{-1} and if $q_1, \dots, q_n$ are the diagonal elements of the commonality matrix Q, then

1. The rank of S is the number of non-zero values in the diagonal of Q.
2. The determinant of S is $q_1 q_2 \cdots q_n$.
3. The trace of S is $q_1 + q_1 + \cdots q_n$.
4. The spectrum of S is $\{q_1, \ldots, q_n\}$, i.e. the commonality numbers are the eigenvalues of S.
5. The linearly independent columns of Q^{-1} are eigenvectors of S, with r_i being associated with the eigenvalue q_i, i.e. $Sr_i = q_i r_i$ for all $i = 1, \ldots n$.

1.1.6 Combining Several Hints

In this section a method for computing the combination of several hints is presented. This method is different from the mere application of the definition of Dempster's rule.

Theorem 10 *For $i = 1, \ldots, k$, let Q_i denote the commonality matrix of a gem-function g_i on the frame Θ. If M denotes the incidence matrix, then*

$$gem(g_1, \ldots, g_k) = M^{-1}(\prod_{i=1}^{k-1} Q_i) M g_k.$$

Proof. The proof of this result can be found in [32]. □

This result can be used to compute the combination of several hints on the same frame Θ.

Corollary 11 *For $i = 1, \ldots, k$, let m_i denote the m-function of a hint $\mathcal{H}_i$ on the frame Θ and let Q_i denote the commonality matrix of m_i. Furthermore, let M denote the incidence matrix. If the combined hint*

$$\mathcal{H} = \mathcal{H}_1 \oplus \ldots \oplus \mathcal{H}_k$$

exists, then its m-function is obtained by normalization of the gem-function

$$g^* = M^{-1}(\prod_{i=1}^{k-1} Q_i) M m_k \tag{1.1.18}$$

Proof. This is a direct consequence of theorem 10. □

If q_k denotes the commonality vector of m_k, then the gem-function g^* in (1.1.18) can also be written as

$$g^* = M^{-1}(\prod_{i=1}^{k-1} Q_i) q_k \tag{1.1.19}$$

because

$$M m_k = q_k.$$

As a special case, this result can be applied to the situation where several copies of the same hint must be combined.

Corollary 12 *Let m denote the m-function of a hint $\mathcal{H}$ on Θ and let Q denote the commonality matrix of m. Furthermore, let M denote the incidence matrix and define $\mathcal{H}_i = \mathcal{H}$ for all $i = 1, \ldots, k$. If the combined hint*

$$\mathcal{H}^* = \mathcal{H}_1 \oplus \ldots \oplus \mathcal{H}_k$$

exists, then its m-function is obtained by normalization of the gem-function

$$g^* = M^{-1} Q^{k-1} M m.$$

Proof. This result is a direct consequence of corollary 11. □

In addition, if q denotes the commonality vector of m, then

$$g^* = M^{-1} Q^{k-1} q$$

according to equation (1.1.19).

1.2 Application

As a simple application of corollary 12, we consider the combination of k copies of a hint $\mathcal{H}$ on a frame $\Theta = \{\theta_1, \theta_2\}$. The ordering of the subsets of Θ is taken to be

$$B_1 = \emptyset,\ B_2 = \{\theta_1\},\ B_3 = \{\theta_2\},\ B_4 = \Theta.$$

The hint $\mathcal{H}$ on Θ is specified by its m-function m given by the column vector

$$\big(0,\, p,\, q,\, r\big)'$$

with $p + q + r = 1$. The commonality vector q of m is

$$q = \big(1,\, p+r,\, q+r,\, r\big)'$$

and the commonality matrix of m is

$$Q = \begin{pmatrix} 1 & 0 & 0 & 0 \\ 0 & p+r & 0 & 0 \\ 0 & 0 & q+r & 0 \\ 0 & 0 & 0 & r \end{pmatrix}.$$

The incidence matrix M is

$$M = \begin{pmatrix} 1 & 1 & 1 & 1 \\ 0 & 1 & 0 & 1 \\ 0 & 0 & 1 & 1 \\ 0 & 0 & 0 & 1 \end{pmatrix},$$

and its inverse matrix is

$$M^{-1} = \begin{pmatrix} 1 & -1 & -1 & 1 \\ 0 & 1 & 0 & -1 \\ 0 & 0 & 1 & -1 \\ 0 & 0 & 0 & 1 \end{pmatrix}.$$

Let $\mathcal{H}^*$ denote the combination of k copies of $\mathcal{H}$ by Dempster's rule of combination, i.e.

$$\mathcal{H}^* = \mathcal{H} \oplus \ldots \oplus \mathcal{H} \qquad (k \text{ terms}).$$

Then the m-function m^* of $\mathcal{H}^*$ is obtained by normalization of the gem-function

$$\begin{aligned} g^* &= M^{-1} Q^{k-1} q \\ &= \left(1 - (p+r)^k - (q+r)^k + r^k,\ (p+r)^k - r^k,\ (q+r)^k - r^k,\ r^k\right)', \end{aligned}$$

which yields

$$m^* = K^{-1}\left(0,\ (p+r)^k - r^k,\ (q+r)^k - r^k,\ r^k\right)' \tag{1.2.20}$$

where

$$K = (p+r)^k + (q+r)^k - r^k. \tag{1.2.21}$$

In particular, if the hint $\mathcal{H}$ is precise, i.e. $r = 0$ and so

$$m = \left(0,\ p,\ q,\ 0\right)',$$

the combined hint $\mathcal{H}^*$ is also precise and its m-function is

$$m^* = K^{-1}\left(0,\ p^k,\ q^k,\ 0\right)' \tag{1.2.22}$$

where

$$K = p^k + q^k. \tag{1.2.23}$$

1.3 The Combination of Closed Hints

In this section we are going to consider the combination of several copies of a hint having an additional special property that will simplify the computation.

Definition 14 *Let $\mathcal{H} = (\Omega, P, \Gamma, \Theta)$ be a hint. Then the set of all focal sets of $\mathcal{H}$ is denoted by $\mathcal{K}^*$ and the set*

$$\mathcal{K} = \mathcal{K}^* \cup \{\emptyset\}$$

is called the kernel of the hint $\mathcal{H}$.

This allows us to introduce the notion of a closed hint as follows.

Definition 15 *A hint $\mathcal{H}$ is called closed if its kernel is closed under intersection, i.e.*

$$K \in \mathcal{K},\ K' \in \mathcal{K} \Rightarrow K \cap K' \in \mathcal{K}.$$

This definition is motivated by the following result.

Theorem 13 *Let $\mathcal{H} = (\Omega, P, \Gamma, \Theta)$ be a closed hint with kernel $\mathcal{K}$. Then $\mathcal{K}$ is also the kernel of the combined hint $\mathcal{H} \oplus \mathcal{H}$.*

Proof. Let us consider two cases. In the first case it is assumed that there exists $K_0 \in \mathcal{K}^*$ and $K'_0 \in \mathcal{K}^*$ such that

$$K_0 \cap K'_0 = \emptyset.$$

Then, by definition of Dempster's rule, the focal sets of $\mathcal{H} \oplus \mathcal{H}$ are the elements of the set

$$\{K \cap K' : K, K' \in \mathcal{K}^*\} - \{\emptyset\}$$

and hence the kernel of $\mathcal{H} \oplus \mathcal{H}$ is the set

$$\mathcal{A} = \{K \cap K' : K, K' \in \mathcal{K}^*\}.$$

So we must prove that $\mathcal{A} = \mathcal{K}$. If $X \in \mathcal{A}$, then $X \in \mathcal{K}$ because $\mathcal{H}$ is closed. Conversely, let $X \in \mathcal{K}$. If $X \neq \emptyset$, then $X \in \mathcal{K}^*$ and $X = X \cap X$, which proves that $X \in \mathcal{A}$. If $X = \emptyset$, then $\emptyset = K_0 \cap K'_0$, which shows that $X \in \mathcal{A}$.

Now let us consider the second case where it is assumed that for all K and K' in $\mathcal{K}^*$, the intersection $K \cap K'$ is not empty. Then the focal sets of $\mathcal{H} \oplus \mathcal{H}$ are the elements of the set

$$\{K \cap K' : K, K' \in \mathcal{K}^*\}$$

and hence the kernel of $\mathcal{H} \oplus \mathcal{H}$ is the set

$$\mathcal{B} = \{K \cap K' : K, K' \in \mathcal{K}^*\} \cup \{\emptyset\}.$$

So we must prove that $\mathcal{B} = \mathcal{K}$. If $X \in \mathcal{B}$, then $X \in \mathcal{K}$ because $\mathcal{H}$ is closed. Conversely, let $X \in \mathcal{K}$. If $X \neq \emptyset$, then $X \in \mathcal{K}^*$ and $X = X \cap X$, which proves that $X \in \mathcal{B}$. If $X = \emptyset$, then obviously $X \in \mathcal{B}$. □

This theorem says that the kernel of a closed hint is invariant by Dempster's rule of combination. From now on, the elements of the kernel of a hint will be denoted by

$$K_1, \ldots, K_n$$

and without loss of generality it is assumed that

$$K_1 = \emptyset \tag{1.3.24}$$

and

$$K_i \subseteq K_j \Rightarrow i \leq j. \tag{1.3.25}$$

The square matrix M of order n corresponding to the ordered elements of the kernel is then defined by

$$M_{ij} = \begin{cases} 1 & \text{if } K_i \subseteq K_j \\ 0 & \text{otherwise} \end{cases} \tag{1.3.26}$$

for all i and j in $\{1, \ldots, n\}$. This matrix is similar to the incidence matrix defined in equation (1.1.15), the only difference is that in the definition (1.3.26) only the elements of the kernel are considered. This considerably reduces the size of the matrix. Note that M is a regular, upper triangular matrix.

Now let m denote the m-function of a closed hint $\mathcal{H}$ with kernel

$$\mathcal{K} = \{K_1, \ldots, K_n\}$$

satisfying (1.3.24) and (1.3.25). Then we define the square matrix S of order n by

$$S_{ij} = \sum \{m(K) : K \in \mathcal{K}, K \cap K_j = K_i\}$$

and, if q denotes the commonality function of $\mathcal{H}$, we define the square matrix Q of order n by

$$Q_{ij} = \begin{cases} q(K_i) & \text{if } i = j \\ 0 & \text{otherwise.} \end{cases}$$

Note that the matrices S and Q are similar to the matrices defined in equations (1.1.14) and (1.1.17), the only difference is that here they are defined with respect to the kernel of $\mathcal{H}$ and not with respect to the entire power set of Θ.

As we are interested in the combination of r copies of the closed $\mathcal{H}$, the following two functions associated with $\mathcal{H}$ are considered.

Definition 16 *Let $\mathcal{H}$ be a closed hint with kernel $\mathcal{K}$ and m-function m. For $r \geq 1$, the τ_r-function of $\mathcal{H}$ is the mapping*

$$\tau_r : \mathcal{K} \longrightarrow [0,1]$$

given by

$$\tau_r(K_i) = \sum \{\prod_{j=1}^{r} m(K_j) : K_j \in \mathcal{K}, \cap_{j=1}^{r} K_j = K_i\}.$$

If $\tau_r(\emptyset) < 1$, then the τ_r^*-function of $\mathcal{H}$ is the mapping

$$\tau_r^* : \mathcal{K}^* \longrightarrow [0,1]$$

given by

$$\tau_r^*(K_i) = \frac{\tau_r(K_i)}{1 - \tau_r(\emptyset)}. \tag{1.3.27}$$

Note that, if $r = 1$, then

$$\tau_1(F) = \tau_1^*(F) = m(F)$$

for every focal set $F \in \mathcal{K}^*$. Furthermore,

$$\sum_{i=1}^{n} \tau_r(K_i) = 1$$

and

$$\sum \{\tau_r^*(F) : F \in \mathcal{K}^*\} = 1.$$

Since the hint $\mathcal{H}$ is closed, theorem 3 implies that the combination of r copies of $\mathcal{H}$ exists if $\tau_r(\emptyset) < 1$, in which case the m-function m_r of the combined hint

$$\mathcal{H}^r = \mathcal{H} \oplus \cdots \oplus \mathcal{H}, \qquad (r \text{ terms})$$

is given by

$$m_r(A) = \begin{cases} \tau_r^*(A) & \text{if } A \in \mathcal{K}^* \\ 0 & \text{otherwise} \end{cases} \tag{1.3.28}$$

for all subsets $A \subseteq \Theta$ because the kernel of $\mathcal{H}^r$ is $\mathcal{K}$ according to theorem 13. Therefore, if we want to determine the m-function of the combination of r copies of the closed hint $\mathcal{H}$, then we need a method for computing its τ_r-function. This function can be specified by the n-dimensional column vector

$$\tau_r = (\tau_r(K_1), \ldots, \tau_r(K_n))'.$$

Theorem 14 *Let m denote the m-function of a closed hint $\mathcal{H}$ with kernel $\mathcal{K}$. Then*

$$\tau_{r+1} = S \cdot \tau_r.$$

Proof. The proof of this result is omitted because it is similar to the proof of theorem 8. □

Theorem 15 *Let m denote the m-function of a closed hint $\mathcal{H}$ with kernel $\mathcal{K}$. Then*

$$S = M^{-1}QM.$$

Proof. The proof of this result is omitted because it is similar to the proof of theorem 9. □

The following result will be useful in the sequel.

Theorem 16 *Let $\mathcal{H}$ be a closed hint with kernel $\mathcal{K}$. If $q_{\mathcal{K}}$ denotes the restriction to $\mathcal{K}$ of the commonality function q of $\mathcal{H}$, then*

$$\tau_r = M^{-1}Q^{r-1}q_{\mathcal{K}}. \tag{1.3.29}$$

Furthermore, if $\tau_r(\emptyset) < 1$, then the combined hint

$$\mathcal{H}^r = \mathcal{H} \oplus \cdots \oplus \mathcal{H} \qquad (r \text{ terms})$$

exists and its m-function m_r is given by

$$m_r(A) = \begin{cases} \tau_r^*(A) & \text{if } A \in \mathcal{K}^* \\ 0 & \text{otherwise.} \end{cases} \tag{1.3.30}$$

Proof. By theorem 14 we have $\tau_r = S^{r-1}\tau_1$ and by theorem 15 we have

$$\tau_r = (M^{-1}QM)^{r-1}\tau_1 = M^{-1}Q^{r-1}M\tau_1.$$

But we obviously have $M\tau_1 = q_{\mathcal{K}}$ and so equation (1.3.29) is proved. The second part of the theorem is simply a repetition of equation (1.3.28). □

1.4 The Theories of Bayes and Fisher

In this section, the theories of Bayes and Fisher will be applied to a small example that we have created to illustrate these two theories for reasoning under uncertainty. This example will be considered again later from the perspective of the theory of functional models.

1.4.1 Jessica's Pregnancy Test

Jessica is a young woman suspecting that she might be pregnant. To find out about her status she goes to the local pharmacy and buys a pregnancy test. The test result indicates that she is not pregnant, but she knows that such tests are not fully trustworthy. The question is then how to evaluate her chances of being pregnant in spite of the negative test result.

1.4.2 The Solution of Bayes

Of course, one possibility is to use the Bayesian theory to analyze the situation. The Bayesian model is as follows. Let θ denote the variable indicating her actual pregnancy status. The set of possible values of θ is $\Theta = \{-1, +1\}$ where -1 means that she is not pregnant and $+1$ means that she is pregnant. Similarly, let ξ denote the variable indicating a test result. The set of possible values of ξ is $X = \{-1, +1\}$ where -1 represents a negative test result and

+1 a positive test result. The reliability of the test can be expressed by two numbers, namely the chance p that the test will indicate a negative result when a woman in not pregnant, and the chance p' that the test will indicate a positive result when a woman is pregnant. It is reasonable to assume that both p and p' are rather high. This information is represented by the conditional probabilities

$$P(\xi = -1|\theta = -1) = p, \ P(\xi = 1|\theta = 1) = p'.$$

Within the Bayesian framework, it is necessary that Jessica provides a prior probability for her being pregnant. Suppose that the prior about her status is given by the probabilities

$$P(\theta = -1) = y, \ P(\theta = +1) = 1 - y.$$

Then, by the theorem of Bayes, the posterior probability that Jessica is not pregnant is

$$P'(\theta = -1) = \frac{yp}{yp + (1-y)(1-p')},$$

which represents Jessica's degree of confidence in the fact that she is not pregnant. Of course, her degree of confidence in the fact that she is pregnant is then

$$P'(\theta = 1) = \frac{(1-y)(1-p')}{yp + (1-y)(1-p')}. \tag{1.4.31}$$

Now assume that the probability of getting a correct negative result is the same as getting a correct positive result, i.e. $p = p'$. This means that the test reveals the true pregnancy status with probability p. This probability is a single-valued indicator of the confidence in the test result. Then by formula (1.4.31) the posterior probability that Jessica is pregnant is

$$P'(\theta = 1) = \frac{(1-y)(1-p)}{yp + (1-y)(1-p)}. \tag{1.4.32}$$

If the uniform prior distribution is used, i.e. $y = 0.5$, then

$$P'(\theta = 1) = \frac{0.5(1-p)}{0.5(p + 1 - p)} = 1 - p. \tag{1.4.33}$$

Formula (1.4.32) is the Bayesian solution to Jessica's pregnancy test problem. For an extensive account of the theoretical foundations and computational methods for reasoning under uncertainty in the Bayesian framework, see Pearl [33].

1.4.3 The Solution of Fisher

It is very possible that Jessica is unable to give the prior probability of her being pregnant because she is not even comfortable with the idea of giving a precise number to estimate this chance. As can be seen in formulas (1.4.31) and (1.4.32), without a prior it is impossible to find the posterior probability of her being pregnant in spite of the negative test result. The Bayesian theory requires prior probabilities to compute posterior probabilities. Although she does not want to give a prior, Jessica is still interested in finding a numerical value expressing her confidence in being pregnant considering the negative test result. How can we find such a numerical value ?

In 1930, R.A. Fisher [17] identified a class of problems in which it appeared to him that inductive probability statements could legitimately be made without prior probabilities being used. It turns out that Jessica's problem when $p = p'$ is one of them. In this example, Fisher's reasoning would go as follows. First define the variable

$$\omega = \theta \cdot \xi \tag{1.4.34}$$

and remark that the distribution of ω does not dependent on θ as the following development shows. Indeed, by equation (1.4.34),

$$\begin{aligned} P(\omega = 1) &= P(\theta = 1, \xi = 1) + P(\theta = -1, \xi = -1) \\ &= P(\xi = 1|\theta = 1)P(\theta = 1) + P(\xi = -1, \theta = -1)P(\theta = -1) \\ &= p \cdot P(\theta = 1) + p \cdot P(\theta = -1) \\ &= p, \end{aligned}$$

and, similarly, it can be shown that $P(\omega = -1) = 1 - p$. Fisher called such a variable ω a *pivotal quantity* or a *pivot*. Generally speaking, a pivotal quantity is a function of the parameter θ and the observation ξ whose distribution is independent of θ. According to equation (1.4.34), the observation $\xi = -1$ Jessica makes does not permit her to infer any information on ω because both $\omega = +1$ and $\omega = -1$ are compatible with the observation. So after observing $\xi = -1$ we still have $P(\omega = 1) = p$ and $P(\omega = -1) = 1 - p$. These probability statements are as valid as before observing $\xi = -1$. The distribution of ω can be seen as a characteristic of the test device. Indeed, $P(\omega = 1) = p$ means that the probability of getting a correct test result is p because $\omega = 1$ whenever θ and ξ indicate the same pregnancy status. Under the assumption that $\omega = 1$, the observation $\xi = -1$ *logically* implies that $\theta = -1$ since $\omega = \theta \cdot \xi$. Since the assumption that $\omega = 1$ is true with probability p, it follows that this probability can be transferred to its logical consequence $\theta = -1$, i.e. $P(\theta = -1) = p$. In a similar way, it can be derived that $P(\theta = 1) = 1 - p$. This argumentation is an instance of *Fisher's fiducial theory*. The probabilities $P(\theta = 1) = 1 - p$ and $P(\theta = -1) = p$ represent the so-called *fiducial distribution of* θ. Fisher noticed that the fiducial distribution coincides with the Bayesian posterior if a uniform prior distribution is assumed for the variable θ (see equation (1.4.33), but here it is assumed that Jessica has no prior distribution, not even the uniform one, and so the statement $P(\theta = +1) = 1 - p$ cannot be justified by Bayes' theorem.

Contrary to this example, there are situations where an observation does provide information about a pivotal quantity $\omega = g(\theta, \xi)$. If Ω denotes the domain of ω, this information is represented by the set Ω' of all ω in Ω that are compatible with the observation. The elements in $\Omega - \Omega'$ are impossible in view of the observation. This knowledge generated by the observation is taken into account by conditioning the initial distribution of ω on the subset Ω'. This idea of conditioning the distribution of the pivot is also present in Fisher's fiducial theory. Using the pivot $\omega = g(\theta, \xi)$ and the observation $\xi = x$, every element $\omega \in \Omega'$ logically implies that $\theta = t$ for some $t \in \Theta$. If P' denotes the conditional distribution of ω given Ω', then the *fiducial probability* of a given t is the probability of all $\omega \in \Omega'$ that logically imply $\theta = t$, i.e.

$$P_{fid}(t) = \sum \{P'(\omega) : \omega \in \Omega',\ \omega \text{ implies } t\}.$$

As we are going to see, the theory of generalized functional models is strongly inspired by this kind of arguments called Fisher's fiducial argument. The paper of Dempster [15] gives an interesting perspective on the relation between the theories of Bayes and Fisher on one hand, and the theory of belief functions on the other hand.

1.5 The Definition and Analysis of a Generalized Functional Model

In this section we develop the theory of generalized functional models using a logic inspired by Fisher's fiducial argument. Within the fiducial theory, the reasoning starts with two variables : the parameter variable θ and the observation variable ξ. Then one tries to find a pivotal quantity $\omega = g(\theta, \xi)$ from which the fiducial distribution of θ is determined.

In contrast, in generalized functional models, we start from three variables : the parameter θ, the observation variable ξ and a third variable ω. The variable ω is a random perturbation whose distribution P is known and assumed to be independent of θ. This is a major difference with the fiducial theory because this theory requires finding a function of θ and ξ whose distribution can be determined and proved to be independent of θ. In the fiducial theory, conditional distributions represent the initial knowledge before any observation is made. However, in generalized functional models the initial knowledge is not given by such conditional distributions, but by the known distribution of ω assumed to be independent of θ, and by a known function of the variables θ and ω to be described below.

The theory of generalized functional models is now formally defined. Let Ω denote the set of all possible values of the random variable ω and as usual let Θ and X denote the set of possible values of θ and ξ respectively. It is assumed that there is exactly one correct but unknown value of θ, which is denoted by θ^*. It is also assumed that Θ, X and Ω are finite. The objective is to generate knowledge about θ^* in the light of the available information. The interaction between the three variables ω, ξ and θ is represented by a function

$$f : \Theta \times \Omega \to X \tag{1.5.35}$$

that is determined from the situation under investigation. Several examples will be considered below. The value $f(\theta, \omega)$ specifies what would inevitably be observed if θ was the correct parameter value and ω was the outcome of the random variable. It is important to remark that the function f is completely determined by the situation under investigation. Thus a generalized functional model is a pair (f, P) where f is a function as in (1.5.35) that is assumed to be known and P is a known probability distribution on Ω that is independent of θ. Figure 1.1 shows that the observed value ξ is a function of the parameter ω and the random variable ω whose probability distribution is P.

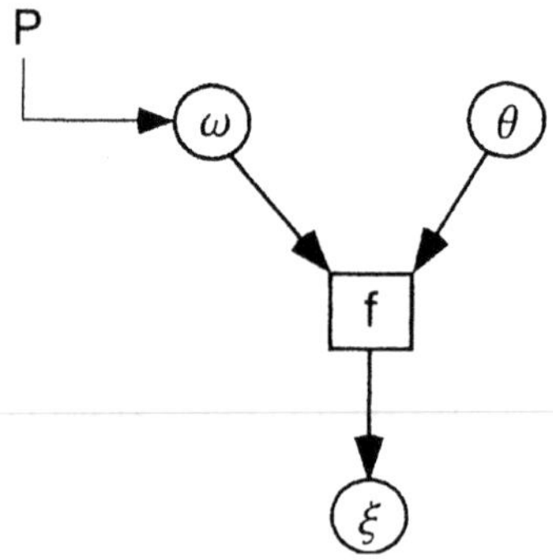

Fig. 1.1. A graphical representation of a generalized functional model

In a generalized functional model (f, P), suppose that the value of the variable ξ is observed to be x, i.e. $\xi = x$. For the time being, assume that this observation is compatible with any value of ω in Ω. Formally, this is expressed by

$$\{\omega \in \Omega : \exists\ \theta \in \Theta \text{ such that } f(\theta, \omega) = x\} = \Omega,$$

which means that the observation x could have been generated by any value ω in Ω. In such a situation it is natural to consider that the distribution P of ω remains the same after observing x, an idea which is also present in the fiducial theory. Under the assumption that the outcome of the random variable was ω, the observation x logically implies that the correct value of θ must be in the set

$$\Gamma_x(\omega) = \{\theta \in \Theta : f(\theta, \omega) = x\}.$$

Therefore, the element ω in Ω is an argument in favor of the hypothesis that θ^* is in $\Gamma_x(\omega)$ because, given the observation x, if ω was the outcome then the hypothesis must be true. Since $P(\omega)$ is the probability of the outcome ω, it follows that ω supports the hypothesis that $\theta^* \in \Gamma_x(\omega)$ to the degree $P(\omega)$. Note that it is possible for $\Gamma_x(\omega)$ to contain more than one element. However, in classical, non generalized, functional models introduced and studied by Fraser [18], Bunke [7], Plante [35], Dawid & Stone [10] and Shafer [39], the sets $\Gamma_x(\omega)$ are always assumed to be singletons, i.e. subsets of Θ containing only one element. So our theory of generalized functional models extends the classical theory of functional models in the sense that the sets $\Gamma_x(\omega)$ need not be singletons.

The reasoning presented above is similar to the one used to define the fiducial distribution of θ when an observation is made. However, in both the fiducial theory and the classical, non generalized, theory of functional models, the sets $\Gamma_x(\omega)$ are always assumed to be singletons, a condition which does not necessarily hold true in our generalized theory of functional models. Figure 1.2 illustrates the direction of the reasoning in a generalized functional model, i.e., given the observation $\xi = x$, assuming ω permits to infer that θ^* is in the subset $\Gamma_x(\omega)$ of Θ.

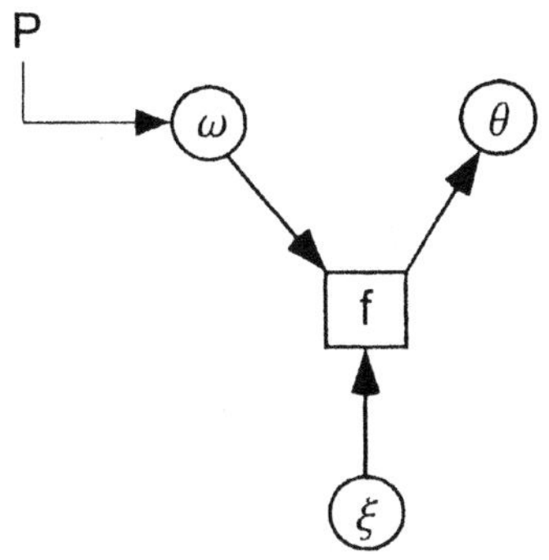

Fig. 1.2. A graphical representation of the reasoning process

Now consider the case where the observation x does provide some knowledge about the random variable ω. This happens when the set

$$v_x(\Theta) = \{\omega \in \Omega : \exists\, \theta \in \Theta \text{ such that } f(\theta, \omega) = x\} \tag{1.5.36}$$

is different from Ω. Only the elements of $v_x(\Theta)$ are possible values for the outcome of the random variable ω in view of the observation x. For this reason, the elements of $v_x(\Theta)$ are said to be compatible with the observation, whereas the elements of $\Omega - v_x(\Theta)$ are said to be incompatible with the observation. Therefore, as in the fiducial theory, the initial probability distribution P of

ω has to be conditioned on $v_x(\Theta)$, which defines a new probability measure P' on $v_x(\Theta)$. Under the assumption that $\omega \in v_x(\Theta)$ was the actual outcome of the random variable, the observation x implies that θ^* is in the set

$$\Gamma_x(\omega) = \{\theta \in \Theta : f(\theta, \omega) = x\}.$$

Since $P'(\omega)$ is the probability of the outcome ω, it follows that ω in $v_x(\Theta)$ supports the hypothesis that θ^* is in $\Gamma_x(\omega)$ to the degree $P'(\omega)$. In other words, ω in $v_x(\Theta)$ is an argument in favor of the hypothesis $\Gamma_x(\omega)$ and $P'(\omega)$ represents the weight of this argument . Finally, note that $\Gamma_x(\omega)$ is not empty for all $\omega \in v_x(\Theta)$.

1.6 Generalized Functional Models and Hints

The goal of this section is to show how the knowledge about θ generated by the observations in a generalized functional model can be expressed by means of the theory of hints, a theory strongly related to the Dempster-Shafer theory of evidence [27, 38]. Given a generalized functional model (f, P), an observation x generates the set $v_x(\Theta)$ of compatible outcomes of the random perturbation. This knowledge is used to condition the probability measure P on $v_x(\Theta)$ and the resulting probability measure is denoted by P_x. By means of the functional relation f and the observation x, each compatible outcome $\omega \in v_x(\Theta)$ logically implies that θ^* is in the subset $\Gamma_x(\omega)$ of Θ. This defines the mapping

$$\Gamma_x : v_x(\Theta) \to 2^\Theta.$$

Since $P_x(\omega)$ is the probability of the outcome ω, it follows that ω supports the hypothesis that $\theta^* \in \Gamma_x(\omega)$ to the degree $P_x(\omega)$. In other words, ω is an argument in favor of the hypothesis $\Gamma_x(\omega)$ and $P_x(\omega)$ is its weight. Putting things together, the observation x and the generalized functional model (f, P) define the hint

$$\mathcal{H}(x) = (v_x(\Theta), P_x, \Gamma_x, \Theta). \tag{1.6.37}$$

It represents the knowledge about θ inferred from the observation x in the generalized functional model. If we define

$$u_x(H) = \{\omega \in v_x(\Theta) : \Gamma_x(\omega) \subseteq H\},$$

then the degree of support of the hypothesis H is

$$sp(H) = P_x(u_x(H))$$

because every element in $u_x(H)$ is an argument of H. The value $sp(H)$ measures the strength of the evidence represented by the observation x that speaks in favor of the hypothesis H. Furthermore, according to equation (1.1.1), for a hypothesis $H \subseteq \Theta$, if we define the set

$$v_x(H) = \{\omega \in v_x(\Theta) : \Gamma_x(\omega) \cap H \neq \emptyset\} \tag{1.6.38}$$

then the degree of plausibility of H is

$$pl(H) = P_x(v_x(H)).$$

The value $pl(H)$ measures the degree of compatibility of the hypothesis H with the evidence provided by the observation x because every element in $v_x(H)$ is compatible with the hypothesis H and the observation x. If we take $H = \Theta$ in equation (1.6.38), then

$$\begin{aligned} v_x(\Theta) &= \{\omega \in v_x(\Theta) : \Gamma_x(\omega) \cap \Theta \neq \emptyset\} \\ &= \{\omega \in v_x(\Theta) : \Gamma_x(\omega) \neq \emptyset\}. \end{aligned} \tag{1.6.39}$$

Since $\Gamma_x(\omega)$ is not empty for all ω in the set $v_x(\Theta)$ defined in equation (1.5.36), the set given in equation (1.6.39) is equal to $v_x(\Theta)$. Therefore, the definition given in equation (1.6.38) is compatible with the definition given in equation (1.5.36).

So far we have considered the case where there is only one observation used to infer about θ. Now consider a second observation x' resulting from another independent realization of the random variable ω. We consider it as the result associated with another random perturbation ω' and we assume that ω and ω' are independent and identically distributed random variables before any observation is made. Let ξ' denote the observation variable associated with the perturbation ω'. Of course, the set of possible values of ξ' and ω' are still X and Ω and the function $f : \Theta \times \Omega \to X$ is still valid when $\xi' = x'$ is observed. In the same way as the observation x generated the hint $\mathcal{H}(x)$, the observation x' generates the hint

$$\mathcal{H}(x') = (v_{x'}(\Theta), P_{x'}, \Gamma_{x'}, \Theta)$$

where $P_{x'}$ is the conditional probability of P given $v_{x'}(\Theta)$. The objective is then to draw inferences about θ given the information represented by the hints $\mathcal{H}(x)$ and $\mathcal{H}(x')$ derived from the observations x and x'.

Under the assumption that the outcome of the two random perturbations was (ω, ω'), the observations x and x' imply that $\theta^* \in \Gamma_x(\omega)$ and $\theta^* \in \Gamma_{x'}(\omega')$ and hence $\theta^* \in \Gamma_x(\omega) \cap \Gamma_{x'}(\omega')$. Therefore, a pair (ω, ω') such that $\Gamma_x(\omega) \cap \Gamma_{x'}(\omega') = \emptyset$ is certainly not a possible pair of outcomes. So, in view of the observations x and x', the set of possible outcomes is

$$\Omega_{x,x'} = \{(\omega, \omega') \in v_x(\Theta) \times v_{x'}(\Theta) : \Gamma_x(\omega) \cap \Gamma_{x'}(\omega') \neq \emptyset\}.$$

Therefore, the product probability measure $P_x P_{x'}$ defined on the product space $v_x(\Theta) \times v_{x'}(\Theta)$ must be conditioned on the event $\Omega_{x,x'}$, which leads to the new probability space $(\Omega_{x,x'}, P_{x,x'})$ where

$$P_{x,x'}(\omega, \omega') = \frac{P_x(\omega) P_{x'}(\omega')}{\sum \{P_x(\omega) P_{x'}(\omega') : (\omega, \omega') \in \Omega_{x,x'}\}}$$

for all (ω, ω') in $\Omega_{x,x'}$. The probability $P_{x,x'}(\omega, \omega')$ expresses the chance of the outcome (ω, ω') after observing x and x'. Putting things together results in the hint

$$\mathcal{H}(x, x') = (\Omega_{x,x'}, P_{x,x'}, \Gamma_{x,x'}, \Theta)$$

where

$$\Gamma_{x,x'}(\omega, \omega') = \Gamma_x(\omega) \cap \Gamma_{x'}(\omega')$$

for all $(\omega, \omega') \in \Omega_{x,x'}$. Of course, $\mathcal{H}(x, x')$ is nothing but the combination of $\mathcal{H}(x)$ and $\mathcal{H}(x')$ by Dempster's rule, i.e.

$$\mathcal{H}(x, x') = \mathcal{H}(x) \oplus \mathcal{H}(x').$$

As usual, degrees of support and degrees of plausibility of hypotheses can be computed from this combined hint. For example, if we define

$$u_{x,x'}(H) = \{(\omega, \omega') \in \Omega_{x,x'} : \Gamma_{x,x'}(\omega, \omega') \subseteq H\},$$

then the degree of support of H induced by the observations x and x' is

$$sp_{x,x'}(H) = P_{x,x'}(u_{x,x'}(H)).$$

When there is a sequence of independent observations $x_1, \ldots, x_n$ in a generalized functional model (f, P), the inferred knowledge about θ is represented by the combined hint

$$\mathcal{H}(x_1, \ldots, x_n) = \oplus_{i=1}^{n} \mathcal{H}(x_i).$$

This hint can then be used to determine degrees of support and degrees of plausibility of hypotheses of interest in the usual way.

1.7 Examples of Generalized Functional Models

In this section we present and analyze three different generalized functional models.

1.7.1 Jessica's Pregnancy Test

The Jessica's pregnancy test problem introduced in section 1.4.1 is analyzed from a functional model perspective. Let θ denote the variable indicating her actual pregnancy status. The set of possible values of θ is $\Theta = \{-1, +1\}$ where -1 means that she is not pregnant and $+1$ means that she is pregnant. Similarly, let ξ denote the variable indicating a test result. The set of possible values of ξ is $X = \{-1, +1\}$ where -1 represents a negative test result and $+1$ a positive test result. Finally, let ω denote the variable indicating whether the pregnancy test is reliable (gives the correct result) or not reliable (gives the opposite of the correct result). The set of possible values of ω is $\Omega = \{-1, +1\}$ where -1 means that the pregnancy test is not reliable and $+1$ means that it is reliable. Furthermore, let r denote the degree of reliability of the pregnancy test, i.e. the test result indicates her true pregnancy status with probability r and it indicates the opposite of her true pregnancy status with probability $1 - r$. Of course, this means that

$$P(\omega = +1) = r, \quad P(\omega = -1) = 1 - r.$$

Now, $\omega = +1$ means that $\xi = \theta$ and $\omega = -1$ means that $\xi = -\theta$. This leads to the function

$$f : \Theta \times \Omega \longrightarrow X$$

given by

$$f(\theta, \omega) = \theta \cdot \omega,$$

i.e.

$$\theta \cdot \omega = \xi.$$

This function f along with the probability r completely specify a functional model representing Jessica's pregnancy test problem. Since the test result is actually negative, i.e $\xi = -1$, it follows that $v_{-1}(\Theta) = \Omega$ and

$$\Gamma_{-1}(1) = \{-1\}, \quad \Gamma_{-1}(-1) = \{1\},$$

which defines the hint

$$\mathcal{H}(-1) = (\Omega, P, \Gamma_{-1}, \Theta).$$

This is a precise hint whose support function is

$$sp(-1) = r \quad sp(+1) = 1 - r.$$

So the degree of support that Jessica is pregnant is $1 - r$ and the degree of support that she is not is r. This is a reasonable result because the reliability of the test device is r and the test says that Jessica is not pregnant.

Similarly, if Jessica's test result is positive, i.e. $\xi = +1$, then $v_{+1}(\Theta) = \Omega$ and

$$\Gamma_{+1}(+1) = \{+1\}, \quad \Gamma_{+1}(-1) = \{-1\},$$

which defines the hint

$$\mathcal{H}(+1) = (\Omega, P, \Gamma_{+1}, \Theta).$$

Now suppose that the test is independently repeated n times and let k denote the number of positive results and $n - k$ the number of negative results. Let $\mathcal{H}'$ denote the hint coming from the combination of k copies of the hint $\mathcal{H}(+1)$ and let $\mathcal{H}''$ denote the hint coming from the combination of

$n-k$ copies of the hint $\mathcal{H}(-1)$. If m' denotes the m-function of $\mathcal{H}'$, then equations (1.2.22) and (1.2.23) imply that

$$m'(1) = \frac{r^k}{r^k + (1-r)^k},$$

and

$$m'(-1) = \frac{(1-r)^k}{r^k + (1-r)^k}.$$

Similarly, if m'' denotes the m-function of $\mathcal{H}''$, then we get

$$m''(1) = \frac{(1-r)^{n-k}}{(1-r)^{n-k} + r^{n-k}},$$

and

$$m''(-1) = \frac{r^{n-k}}{(1-r)^{n-k} + r^{n-k}}.$$

Now let

$$\mathcal{H} = \mathcal{H}' \oplus \mathcal{H}''$$

denote the hint corresponding to the combination of all positive and negative test results. Using theorem 3, it is then easy to verify that the support function of $\mathcal{H}$ is

$$sp(-1) = \frac{(1-r)^k r^{n-k}}{r^k(1-r)^{n-k} + (1-r)^k r^{n-k}}$$

and

$$sp(+1) = \frac{r^k(1-r)^{n-k}}{r^k(1-r)^{n-k} + (1-r)^k r^{n-k}}.$$

1.7.2 Policy Identification (I)

In this example, the following situation is considered. Peter is in room A and has a regular coin in front of him. He flips the coin and observes what shows up : either H or T (*heads* or *tails*). Then he decides between the following two policies : either tell Paul, who is in room B, what actually showed up on the coin (policy 1) or tell him that the coin showed heads up, regardless of what actually showed up (policy 2). Of course, Paul is unaware of the policy that Peter has chosen. Peter flips the coin n times in total and each time reports to Paul according to the policy he has chosen. Paul knows that Peter is using the same policy each time the coin is flipped. From the sequence of heads and possibly tails that he receives from Peter, Paul wants to infer information about the policy Peter is using. Is it policy 1 or policy 2 ?

To answer this question, Paul builds a generalized functional model of the situation. The parameter space is $\Theta = \{\theta_1, \theta_2\}$ where θ_1 means that Peter is using policy 1, and θ_2 that he is using policy 2. The observation space is $X = \{H, T\}$ (the set of possible reports to Paul after one flip of the coin) and the domain of ω is $\Omega = \{\omega_1, \omega_2\}$, where ω_1 means that heads shows up when the coin is flipped, and ω_2 means that tails shows up when the coin is flipped. Since the coin is a regular one, we have $P(\omega_1) = P(\omega_2) = 0.5$. In this example, as the following reasoning shows, the value of the observation in X is uniquely determined by the value of the parameter in Θ and the value of the variable ω. Let

$$f : \Theta \times \Omega \longrightarrow X$$

denote this functional correspondance and let us start with the value of $f(\theta_1, \omega_1)$. It is clear that under policy 1, if heads shows up then Peter will report H to Paul, which means that $f(\theta_1, \omega_1) = H$. It can easily be seen that the complete function f is given by

$$\begin{array}{ll} f(\theta_1, \omega_1) = H & f(\theta_1, \omega_2) = T \\ f(\theta_2, \omega_1) = H & f(\theta_2, \omega_2) = H. \end{array}$$

First let us analyze the situation when only one observation is reported to Paul. If H is reported, then by the equation (1.5.36) we have

$$v_H(\Theta) = \{\omega \in \Omega : \exists\, \theta \in \Theta \text{ such that } f(\theta, \omega) = H\} = \Omega$$

and so the information that can be inferred on Θ is represented by the hint

$$\mathcal{H}(H) = (\Omega, P, \Gamma_H, \Theta)$$

where

$$\Gamma_H(\omega_2) = \{\theta_2\}, \qquad \Gamma_H(\omega_1) = \Theta.$$

It is important to remark that this hint is not precise because $\Gamma_H(\omega_1)$ is not a singleton. Indeed, $\Gamma_H(\omega_1) = \Theta$ because if H is reported and the coin shows heads up then Peter could have used either policy 1 or policy 2. On the other hand, $\Gamma_H(\omega_2) = \{\theta_2\}$ because if H is reported and the coin shows tails up then Peter is undoubtedly using policy 2. The hypothesis that Peter is using policy 2 is supported with strength $P(\omega_2) = 0.5$ by the observation H, whereas there is no support in favor of the fact that Peter is using policy 1. So the support function associated with $\mathcal{H}(H)$ is given by

$$sp_H(\theta_1) = 0, \qquad sp_H(\theta_2) = 0.5.$$

Since $\mathcal{H}(H)$ is not precise, this support function is not a regular probability measure on Θ because $sp_H(\theta_1) + sp_H(\theta_2) < 1$. The plausibility function associated with $\mathcal{H}(H)$ is given by

$$pl_H(\theta_1) = 0.5, \qquad pl_H(\theta_2) = 1.$$

As explained in section 1.1.1, this means that the degree of compatibility between the hypothesis that Peter is using policy 1 and the observation H

is 0.5, whereas nothing is speaking against the hypothesis that he is using policy 2, it is fully compatible with the observation H.

If T is reported, then

$$v_T(\Theta) = \{\omega \in \Omega : \exists\ \theta \in \Theta \text{ such that } f(\theta,\omega) = T\} = \{\omega_2\}.$$

This implies that we have to condition P on $\{\omega_2\}$, which results in the deterministic probability measure $P'(\omega_2) = 1$. Since $\Gamma_T(\omega_2) = \{\theta_1\}$, which means that if T is reported and the coin turned tails then Peter is using policy 1, it follows that the hint derived from the observation T is

$$\mathcal{H}(T) = (\{\omega_2\}, P', \Gamma_T, \Theta).$$

Then of course the corresponding support function is

$$sp_T(\theta_1) = 1, \qquad sp_T(\theta_2) = 0,$$

which means that Peter is definitely using policy 1. The hint $\mathcal{H}(T)$ is called a deterministic hint on θ_1 because it is certain that θ_1 is the correct value of the parameter.

Now let us analyze the situation where n reports are made. Let k denote the number of heads reported and $n-k$ the number of tails reported. The hint on θ that can be derived from this collection of observations is

$$\mathcal{H}(n,k) = (\oplus_{i=1}^{k}\mathcal{H}(H)) \ \oplus \ (\oplus_{i=1}^{n-k}\mathcal{H}(T)).$$

If at least one tails is reported (i.e. $n-k > 0$), then it can easily be proved that $\mathcal{H}(n,k)$ is again a deterministic hint on θ_1, which means that Peter is using policy 1. This is of course the correct result because if at least one tails is reported then Peter cannot be using policy 2.

Now let us examine the situation where no tails is reported. Using equations (1.2.20) and (1.2.21) with $p = 0$, $q = 0.5$ and $r = 0.5$, it can easily be shown that the m-function of the non-precise hint $\mathcal{H}(n,n)$ is

$$m(\theta_1) = 0, \quad m(\theta_2) = 1 - 0.5^n, \quad m(\Theta) = 0.5^n.$$

In consequence, we have

$$sp(\theta_1) = 0 \qquad sp(\theta_2) = 1 - 0.5^n$$

and

$$pl(\theta_1) = 0.5^n \qquad pl(\theta_2) = 1.$$

The hypothesis θ_1 is becoming less and less plausible as the number of heads reported increases, which is of course very reasonable.

1.7.3 Policy Identification (II)

In this example we consider the following situation. Peter is in room A and has two coins in front of him, one red and one blue. Peter tells Paul, who is in room B, that when the red coin is flipped, the probability that heads shows up is p_1 and when the blue coin is flipped, the probability that heads shows up is p_2. Peter successively flips the red and blue coins and observes what shows up : either H or T for each coin. Then he decides between the following two policies: either tell Paul what showed up on the red coin (policy 1) or tell him what showed up on the blue coin (policy 2). After the policy is chosen, he informs Paul about what showed up on the coin specified by the policy. Of course, Paul is unaware of the policy that Peter is using. This experiment is repeated n times in total, thereby assuming that the policy Peter is using remains the same throughout all experiments. From the sequence of heads and tails that he receives from Peter, Paul wants to infer information about the policy Peter is using, policy 1 or policy 2 ?

To answer this question, Paul decides to build a generalized functional model of the situation. The parameter space is $\Theta = \{\theta_1, \theta_2\}$ where θ_1 means that Peter is using policy 1, and θ_2 that he is using policy 2. The observation space is $X = \{H, T\}$. The joint domain of the random experiment is constructed from the two random variables involved in the experiment, namely the result of flipping the red coin (random variable ω) and the result of flipping the blue coin (random variable ω'). More precisely, the domain of ω is $\Omega = \{\omega_1, \omega_2\}$ where ω_1 means that the red coin shows heads up and ω_2 that it shows tails up. By analogy, we define the domain $\Omega' = \{\omega'_1, \omega'_2\}$ of the random variable ω'. The joint domain of (ω, ω') is then $\Omega \times \Omega'$ and the corresponding probabilities are

$$P(\omega_1, \omega'_1) = p_1 p_2, \qquad P(\omega_1, \omega'_2) = p_1(1 - p_2)$$

$$P(\omega_2, \omega'_1) = (1 - p_1)p_2, \qquad P(\omega_2, \omega'_2) = (1 - p_1)(1 - p_2).$$

In this situation, there is again a unique function

$$f : \Theta \times \Omega \times \Omega' \longrightarrow X$$

specifying the value of the observation in X assuming a parameter value in Θ and an outcome in $\Omega \times \Omega'$. Let us start with the value of $f(\theta_1, \omega_1, \omega'_1)$. It is clear that under policy 1, if the red coin shows heads up then Peter will report H to Paul, which implies that $f(\theta_1, \omega_1, \omega'_1) = H$ and $f(\theta_1, \omega_1, \omega'_2) = H$. It can easily be seen that the complete function f is given by

$$\begin{array}{ll} f(\theta_1, \omega_1, \omega'_1) = H & f(\theta_1, \omega_1, \omega'_2) = H \\ f(\theta_1, \omega_2, \omega'_1) = T & f(\theta_1, \omega_2, \omega'_2) = T \\ f(\theta_2, \omega_1, \omega'_1) = H & f(\theta_2, \omega_1, \omega'_2) = T \\ f(\theta_2, \omega_2, \omega'_1) = H & f(\theta_2, \omega_2, \omega'_2) = T. \end{array}$$

If H is reported, then

$$v_H(\Theta) = \{(\omega_1, \omega_1'), (\omega_1, \omega_2'), (\omega_2, \omega_1')\}$$

and the conditional distribution of P on $v_H(\Theta)$ is the probability P' given by

$$P'((\omega_1, \omega_1')) = \frac{p_1 p_2}{p_1 + p_2 - p_1 p_2} \qquad P'((\omega_1, \omega_2')) = \frac{p_1(1-p_2)}{p_1 + p_2 - p_1 p_2}$$

and

$$P'((\omega_2, \omega_1')) = \frac{(1-p_1)p_2}{p_1 + p_2 - p_1 p_2}.$$

Also, we have

$$\Gamma_H((\omega_1, \omega_1')) = \Theta, \; \Gamma_H((\omega_1, \omega_2')) = \{\theta_1\}, \; \Gamma_H((\omega_2, \omega_1')) = \{\theta_2\},$$

which shows that the hint

$$\mathcal{H}(H) = (v_H(\Theta), P', \Gamma_H, \Theta)$$

is not precise. Its corresponding support function is given by

$$sp_H(\theta_1) = \frac{p_1(1-p_2)}{p_1 + p_2 - p_1 p_2} \qquad sp_H(\theta_2) = \frac{(1-p_1)p_2}{p_1 + p_2 - p_1 p_2}.$$

On the other hand, if T is reported, then

$$v_T(\Theta) = \{(\omega_1, \omega_2'), (\omega_2, \omega_1'), (\omega_2, \omega_2')\}$$

and the conditional distribution of P on $v_T(\Theta)$ is the probability P' given by

$$P'((\omega_1, \omega_2')) = \frac{p_1(1-p_2)}{1 - p_1 p_2} \qquad P'((\omega_2, \omega_1')) = \frac{(1-p_1)p_2}{1 - p_1 p_2}$$

and

$$P'((\omega_2, \omega_2')) = \frac{(1-p_1)(1-p_2)}{1 - p_1 p_2}.$$

Also, we have

$$\Gamma_T((\omega_1, \omega_2')) = \{\theta_2\}, \; \Gamma_T((\omega_2, \omega_1')) = \{\theta_1\}, \; \Gamma_T((\omega_2, \omega_2')) = \Theta,$$

which shows that the hint

$$\mathcal{H}(T) = (v_T(\Theta), P', \Gamma_T, \Theta)$$

is also not precise. Its corresponding support function is given by

$$sp_T(\theta_1) = \frac{(1-p_1)p_2}{1 - p_1 p_2} \qquad sp_T(\theta_2) = \frac{p_1(1-p_2)}{1 - p_1 p_2}.$$

Now let us analyze the situation where n reports are made, k of which being H and $n-k$ being T. Then the hint on Θ induced by this collection of observations is

$$\mathcal{H}(n,k) = (\oplus_{i=1}^{k}\mathcal{H}(H)) \oplus (\oplus_{i=1}^{n-k}\mathcal{H}(T)).$$

The m-function of $\oplus_{i=1}^{k}\mathcal{H}(H)$ and the m-function of $\oplus_{i=1}^{n-k}\mathcal{H}(T)$ can easily be obtained by equations (1.2.20) and (1.2.21), which in turn can be used to compute the m-function of $\mathcal{H}(n,k)$ by a simple application of theorem 3. If m denotes the m-function of $\mathcal{H}(n,k)$, we get

$$m(\theta_1) = \frac{p_1^k(1-p_1)^{n-k} - (p_1p_2)^k((1-p_1)(1-p_2))^{n-k}}{D},$$

$$m(\theta_2) = \frac{p_2^k(1-p_2)^{n-k} - (p_1p_2)^k((1-p_1)(1-p_2))^{n-k}}{D}$$

and

$$m(\Theta) = \frac{(p_1p_2)^k((1-p_1)(1-p_2))^{n-k}}{D}$$

where

$$D = p_1^k(1-p_1)^{n-k} + p_2^k(1-p_2)^{n-k} - (p_1p_2)^k((1-p_1)(1-p_2))^{n-k}.$$

From this result, the support and plausibility functions of $\mathcal{H}(n,k)$ can be easily derived.

1.8 Prior Information

In a situation where the initial knowledge about the unknown parameter contains a prior probability distribution P_0 on Θ, it is easy to include this information into a generalized functional model (f, P). Indeed, such a prior information can be viewed as a precise hint on Θ given by

$$\mathcal{H}_0 = (\Theta, P_0, \Gamma, \Theta)$$

where $\Gamma(\theta) = \{\theta\}$ for all $\theta \in \Theta$. The support and plausibility functions of this hint is nothing but the prior probability distribution P_0. Then this hint $\mathcal{H}_0$ is combined with the hint $\mathcal{H}(x_1,\ldots,x_n)$ on Θ coming from the observations $x_1,\ldots,x_n$. This results in a new hint

$$\mathcal{H} = \mathcal{H}_0 \oplus \mathcal{H}(x_1,\ldots,x_n)$$

expressing the knowledge about Θ derived from the complete information about the problem under investigation. Then, as usual, degrees of support and degrees of plausibility can be determined from the hint $\mathcal{H}$. It turns out that $\mathcal{H}$ is a precise hint because the combination of a precise hint with any other hint is always again a precise hint.

On the other hand, from a generalized functional model (f, P), the following conditional distributions of ξ can be defined:

$$Pr(\xi = x|\theta) = P(\{\omega \in \Omega : f(\theta, \omega) = x\}).$$

Such conditional probability distributions specify what is called a *distribution model* [10]. These conditional distributions, together with a prior probability distribution $\tilde{P}$ on Θ, completely specify a Bayesian model which is denoted by $(Pr, \tilde{P})$.

Theorem 17 *Let $\mathcal{H}(x_1, \ldots, x_n)$ denote the hint on Θ inferred from the observations $x_1, \ldots, x_n$ in a generalized functional model (f, P) with associated distribution model Pr. Let $\mathcal{H}_0$ be the precise hint representing the prior distribution P_0 on Θ and let sp denote the support function of the precise hint*

$$\mathcal{H} = \mathcal{H}_0 \oplus \mathcal{H}(x_1, \ldots, x_n).$$

Furthermore, let P_{post} denote posterior distribution on Θ given the observations $x_1, \ldots, x_n$ in the Bayesian model (Pr, P_0) specified by the conditional distributions Pr and the prior distribution P_0. Then

$$sp(\theta) = P_{post}(\theta) = \frac{P_0(\theta) \prod_{i=1}^{n} Pr(\xi = x_i|\theta)}{\sum_{\theta \in \Theta} P_0(\theta) \prod_{i=1}^{n} Pr(\xi = x_i|\theta)} \tag{1.8.40}$$

for all $\theta \in \Theta$ and so the support function of the hint $\mathcal{H}$ inferred from the generalized functional model coincides with the posterior distribution inferred from the Bayesian model.

Proof. Let $\mathcal{H}(x_i)$ denote the hint on Θ inferred from the observation x_i in the generalized functional model (f, P), i.e.

$$\mathcal{H}(x_i) = (v_{x_i}(\Theta), P'_{x_i}, \Gamma_{x_i}, \Theta)$$

where

$$v_{x_i}(\Theta) = \{\omega \in \Omega : \exists\, \theta \in \Theta \text{ such that } f(\theta, \omega) = x_i\},$$

and, for all $\omega \in v_{x_i}(\Theta)$,

$$P'_{x_i}(\omega) = \frac{P(\omega)}{P(v_{x_i}(\Theta))} \tag{1.8.41}$$

and

$$\Gamma_{x_i}(\omega) = \{\theta \in \Theta : f(\theta, \omega) = x_i\}.$$

Remark that we have

$$\mathcal{H}(x_1, \ldots, x_n) = \oplus_{i=1}^{n} \mathcal{H}(x_i). \tag{1.8.42}$$

According to theorem 5, the hint $\mathcal{H}$ is precise and its m-function is given by

$$m(\theta) = \frac{P_0(\theta) pl_{x_1,\ldots,x_n}(\theta)}{\sum_{\theta \in \Theta} P_0(\theta) pl_{x_1,\ldots,x_n}(\theta)} \tag{1.8.43}$$

where $pl_{x_1,\ldots,x_n}$ denotes the plausibility function of $\mathcal{H}(x_1, \ldots, x_n)$. By equation (1.8.42) and theorem 4, there exits a constant C such that

$$q_{x_1,\ldots,x_n}(\theta) = C \cdot \prod_{i=1}^{n} q_{x_i}(\theta),$$

where $q_{x_1,\ldots,x_n}$ and q_{x_i} denote the commonality function of $\mathcal{H}(x_1, \ldots, x_n)$ and $\mathcal{H}(x_i)$ respectively. Since, for every hint, the plausibility function and the commonality functions coincide on singletons, we have

$$pl_{x_1,\ldots,x_n}(\theta) = C \cdot \prod_{i=1}^{n} pl_{x_i}(\theta). \tag{1.8.44}$$

The equations

$$\begin{aligned} v_{x_i}(\{\theta\}) &= \{\omega \in v_{x_i}(\Theta) : \Gamma_{x_i}(\omega) \cap \{\theta\} \neq \emptyset\} \\ &= \{\omega \in v_{x_i}(\Theta) : \theta \in \Gamma_{x_i}(\omega)\} = \{\omega \in v_{x_i}(\Theta) : f(\theta, \omega) = x_i\} \end{aligned}$$

imply that

$$\begin{aligned} pl_{x_i}(\theta) &= P'_{x_i}(v_{x_i}(\{\theta\}) \\ &= P'_{x_i}(\{\omega \in v_{x_i}(\Theta) : f(\theta, \omega) = x_i\}) \\ &= \sum \{P'_{x_i}(\omega) : \omega \in v_{x_i}(\Theta), f(\theta, \omega) = x_i\}. \end{aligned}$$

But since, for every θ in Θ,

$$\{\omega \in v_{x_i}(\Theta) : f(\theta, \omega) = x_i\} = \{\omega \in \Omega : f(\theta, \omega) = x_i\},$$

we can write, according to equation (1.8.41),

$$\begin{aligned} pl_{x_i}(\theta) &= \sum \{\frac{P(\omega)}{P(v_{x_i}(\Theta))} : \omega \in v_{x_i}(\Theta), f(\theta, \omega) = x_i\} \\ &= \sum \{\frac{P(\omega)}{P(v_{x_i}(\Theta))} : \omega \in \Omega, f(\theta, \omega) = x_i\} \\ &= \frac{1}{P(v_{x_i}(\Theta))} \sum \{P(\omega) : \omega \in \Omega, f(\theta, \omega) = x_i\} \\ &= \frac{1}{P(v_{x_i}(\Theta))} Pr(\xi = x_i | \theta). \end{aligned}$$

So by equation (1.8.44) we conclude that there is a constant K such that

$$pl_{x_1,\ldots,x_n}(\theta) = K \cdot \prod_{i=1}^{n} Pr(\xi = x_i | \theta) \tag{1.8.45}$$

and hence, by equation (1.8.43),

$$m(\theta) = \frac{P_0(\theta) \prod_{i=1}^{n} Pr(\xi = x_i|\theta)}{\sum_{\theta \in \Theta} P_0(\theta) \prod_{i=1}^{n} Pr(\xi = x_i|\theta)}. \qquad (1.8.46)$$

On the other hand, with respect to the Bayesian model (Pr, P_0), the posterior distribution on Θ given the observations $x_1, \ldots, x_n$ is

$$P_{post}(\theta|x_1, \ldots, x_n) = \frac{P_0(\theta) \prod_{i=1}^{n} Pr(\xi = x_i|\theta)}{\sum_{\theta \in \Theta} P_0(\theta) \prod_{i=1}^{n} Pr(\xi = x_i|\theta)}. \qquad (1.8.47)$$

Since $sp(\theta) = m(\theta)$ because $\mathcal{H}$ is precise, equations (1.8.46) and (1.8.47) imply equation (1.8.40), which proves the theorem. □

Theorem 18 *Let $sp_{x_1,\ldots,x_n}$ denote the support function of the hint*

$$\mathcal{H}(x_1, \ldots, x_n)$$

on Θ inferred from the observations $x_1, \ldots, x_n$ in a generalized functional model (f, P) with associated distribution model Pr. Let P_{post} denote the posterior distribution on Θ given the observations $x_1, \ldots, x_n$ in the Bayesian model (Pr, P^u) specified by the conditional distributions Pr and the prior uniform distribution P^u. If $\mathcal{H}(x_1, \ldots, x_n)$ is precise then

$$sp_{x_1,\ldots,x_n}(\theta) = P_{post}(\theta) = \frac{\prod_{i=1}^{n} Pr(\xi = x_i|\theta)}{\sum_{\theta \in \Theta} \prod_{i=1}^{n} Pr(\xi = x_i|\theta)}$$

for all $\theta \in \Theta$ and so the support function of the hint $\mathcal{H}(x_1, \ldots, x_n)$ inferred from the generalized functional model coincides with the posterior distribution inferred from the Bayesian model (Pr, P^u).

Proof. Let $m_{x_1,\ldots,x_n}$ denote the m-function of the precise hint $\mathcal{H}(x_1, \ldots, x_n)$. If $\mathcal{H}^u$ denotes the precise hint corresponding to the prior uniform probability distribution P^u, then of course the m-function m^u of $\mathcal{H}^u$ satisfies

$$m^u(\theta) = P^u(\theta)$$

for all $\theta \in \Theta$. If m denotes the m-function of the precise hint

$$\mathcal{H} = \mathcal{H}^u \oplus \mathcal{H}(x_1, \ldots, x_n),$$

then theorem 6 implies that

$$m(\theta) = \frac{P^u(\theta) m_{x_1,\ldots,x_n}(\theta)}{\sum_{\theta \in \Theta} P^u(\theta) m_{x_1,\ldots,x_n}(\theta)}.$$

Therefore, since $P^u(\theta)$ is a constant, we conclude that

$$m(\theta) = \frac{m_{x_1,\ldots,x_n}(\theta)}{\sum_{\theta \in \Theta} m_{x_1,\ldots,x_n}(\theta)} = m_{x_1,\ldots,x_n}(\theta) \tag{1.8.48}$$

because $\mathcal{H}(x_1, \ldots, x_n)$ is precise. Since $\mathcal{H}$ is also precise, equation (1.8.48) implies that

$$\mathcal{H} = \mathcal{H}(x_1, \ldots, x_n).$$

If sp denotes the support function of $\mathcal{H}$, then this equation shows that

$$sp(\theta) = sp_{x_1,\ldots,x_n}(\theta) \tag{1.8.49}$$

for all $\theta \in \Theta$. But by theorem 17 we have

$$\begin{aligned} sp(\theta) &= P_{post}(\theta) \\ &= \frac{P^u(\theta) \prod_{i=1}^n Pr(\xi = x_i|\theta)}{\sum_{\theta \in \Theta} P^u(\theta) \prod_{i=1}^n Pr(\xi = x_i|\theta)} \\ &= \frac{\prod_{i=1}^n Pr(\xi = x_i|\theta)}{\sum_{\theta \in \Theta} \prod_{i=1}^n Pr(\xi = x_i|\theta)}, \end{aligned}$$

which, along with equation (1.8.49), completes the proof of the theorem. □

2. The Plausibility and Likelihood Functions

The notion of likelihood is an important concept in modern statistics. In particular, the likelihood ratio has been used by several authors [19, 37] to measure the strength of the evidence represented by observations in statistical problems. This idea works fine when the goal is to evaluate the strength of the available evidence for a simple hypothesis versus another simple hypothesis. However, the applicability of this idea is limited to simple hypotheses because the likelihood function is primarily defined on points, i.e. simple hypotheses, of the parameter space. In this chapter, we extend the notion of likelihood from simple to composite hypotheses. It turns out that this extended notion of likelihood is the plausibility function of the hint derived from a generalized functional model. This allows us to define a general weight of evidence that is applicable to both simple and composite hypotheses. Classical distribution models introduced in the previous chapter do not convey enough information to generate a natural and general concept of weight of evidence. In other words, they are too coarse a representation of the mechanical process underlying the generation of the data observed. Generalized functional models and the theory of hints are the appropriate tools for defining such a general concept. Furthermore, a weight of evidence can be given a concrete significance by finding a well understood and simple situation leading to the same weight of evidence. The relevant concepts and ideas are explained by means of a familiar urn problem in addition to the general analysis of a real-world medical problem.

2.1 The Likelihood Ratio as a Weight of Evidence

In a recent book, Richard Royall [37] explains that the likelihood ratio is the appropriate concept for measuring the strength of the evidence represented by observations in statistical problems. He gives several arguments in favor of this idea and shows its relevance in the field of testing statistical hypotheses. This interpretation of likelihood is based on Ian Hacking's *law of likelihood* [19] :

If hypothesis A implies that the probability that a random variable X takes the value x is $P_A(x)$, while hypothesis B implies that the probability is $P_B(x)$, then the observation $X = x$ is evidence supporting A over B if and only if

$P_A(x) > P_B(x)$, and the likelihood ratio, $P_A(x)/P_B(x)$, measures the strength of that evidence (I. Hacking, 1965).

To elaborate on this idea we consider two spaces : the parameter space Θ and the observation space X. The set Θ is the set of possible values of the parameter variable θ, whereas X is the set of possible values of the observable variable ξ. There is exactly one correct but unknown value of the parameter variable θ, which is denoted by θ^*. As we have seen in the previous chapter and following Dawid & Stone [10], a distribution model is the specification of conditional distributions P_θ of the variable ξ given θ, i.e.

$$P(\xi = x|\theta) = P_\theta(x).$$

For each value x in X, the function $l_x : \Theta \to [0, 1]$ given by $l_x(\theta) = P_\theta(x)$ is the likelihood function of the observation x. According to the law of likelihood, the observation $X = x$ is evidence supporting the hypothesis $\{\theta\}$ over the hypothesis $\{\theta'\}$ if

$$l_x(\theta)/l_x(\theta') > 1$$

and the weight of that evidence is precisely

$$w_x(\theta, \theta') = l_x(\theta)/l_x(\theta'). \tag{2.1.1}$$

This defines a function

$$w_x : \Theta \times \Theta \to [0, \infty]$$

which is called the *weight of evidence function.*

Given a distribution model and an observation x, we can compute the weight of evidence of a simple hypothesis $H = \{\theta\}$ over a simple hypothesis $H' = \{\theta'\}$. However, the weight of evidence is not defined for composite hypotheses (a hypothesis is composite if it contains more than one element). As Royall [37] says, in general the law of likelihood is silent for composite hypotheses. In the next section we are going to define a general notion of weight of evidence that is applicable to any kind of hypotheses. Of course, this general weight of evidence will be compatible with the likelihood ratio, i.e. when the general weight of evidence is applied to simple hypotheses the result is the likelihood ratio.

2.2 The Weight of Evidence for Composite Hypotheses

In the previous chapter, the classical notion of a functional model introduced by Bunke [7] and studied by Dawid & Stone [10] has been generalized to so-called generalized functional models, abbreviated by GFM, by allowing the sets $\Gamma_x(\omega)$ to contain more than one element. It turns out that these models can be used to give a sound and reasonable definition of a general weight of evidence. Distribution models do not contain enough information to be able to induce weights of evidence for composite hypotheses. We need to leave the realm of distribution models and enter the richer class of generalized functional models to reach that goal.

As we have seen in the previous chapter, a generalized functional model is a pair (f, P) where f is a function and P is a probability distribution. If Ω denotes the set of possible values of a random perturbation (or random source), then the function f is a mapping $f : \Theta \times \Omega \to X$ with $f(\theta, \omega)$ representing the value of the variable ξ that must necessarily be observed if θ is the correct value of the parameter and the realization of the random perturbation is ω. The distribution of the random perturbation is the known probability measure P on Ω, and the function f is assumed to be known. Note in passing that a GFM can be seen as a parameterized particular causal theory in the sense of Pearl [34]. The observation of the value x for the variable ξ in a GFM generates some information about the unknown correct value of the parameter. As we have seen in the previous chapter, this information is a hint which is denoted by $\mathcal{H}(x)$.

From a GFM we can logically derive a unique distribution model as follows. For every $\theta \in \Theta$, let P_θ be the probability distribution on X given by

$$P_\theta(x) = P(\{\omega \in \Omega : f(\theta, \omega) = x\}).$$

The plausibility function pl_x associated with the hint $\mathcal{H}(x)$ has a very interesting property. Indeed, if l_x denotes the likelihood function of the distribution model associated with the GFM when x is observed, then it can easily be verified that

$$pl_x(\theta) = c \cdot l_x(\theta)$$

for all $\theta \in \Theta$, where c is a positive constant that does not depend on θ. This means that the plausibility function of $\mathcal{H}(x)$ restricted to simple hypotheses is proportional to the likelihood function of the associated distributional model. From this result, given an observation x, the following definition is quite natural :

Definition 17 *Given an observation x, the weight of evidence of a general hypothesis $H \subseteq \Theta$ over a general hypothesis $H' \subseteq \Theta$ is given by*

$$w_x(H, H') = pl_x(H)/pl_x(H').$$

The weight of evidence of the simple hypothesis $H = \{\theta\}$ over the simple hypothesis $H' = \{\theta'\}$ is then

$$pl_x(\theta)/pl_x(\theta') = (c \cdot l_x(\theta))/(c \cdot l_x(\theta')) = l_x(\theta)/l_x(\theta'),$$

which shows that the definition 17 extends to composite hypotheses the weight of evidence for simple hypotheses given in equation (2.1.1).

Of course, the concepts defined above for a single observation can easily be extended to the situation of several independent observations, i.e. observations resulting from stochastically independent realizations of the random perturbation.

First consider the weight of evidence in the context of distribution models. For all $\theta \in \Theta$, let P_θ be a distribution on X. In a total of r independent observations of the variable ξ, suppose that the value x is observed m times and the value x' is observed n times $(m+n=r)$. The case where more than two different observations of ξ are made can be treated in a similar way. These r observations are considered as independent realizations of the variable ξ. Given these observations, the likelihood of the parameter value θ is

$$l_{m,n}(\theta) = \prod_{i=1}^{m} l_x(\theta) \cdot \prod_{i=1}^{n} l_{x'}(\theta).$$

Then the law of likelihood implies that the weight of evidence of θ over θ' is

$$w_{m,n}(\theta,\theta') = l_{m,n}(\theta)/l_{m,n}(\theta') = \prod_{i=1}^{m} w_x(\theta,\theta') \cdot \prod_{i=1}^{n} w_{x'}(\theta,\theta'). \tag{2.2.2}$$

However, the weight of evidence for composite hypotheses is not defined in the context of distribution models.

Now consider the same observations from the point of view of a GFM having the given distributions P_θ as its associated distribution model. Since the observations are independent, it is reasonable to use Dempster's rule of combination to combine the hints coming from each observation. The resulting hint on Θ, which is denoted by $\mathcal{H}_{m,n}$, represents the information on Θ based on all observations :

$$\mathcal{H}_{m,n} = (\oplus_{i=1}^{m}\mathcal{H}(x)) \oplus (\oplus_{i=1}^{n}\mathcal{H}(x')).$$

If $pl_{m,n}$ denotes the plausibility function of $\mathcal{H}_{m,n}$, then the following definition is quite natural.

Definition 18 *The weight of evidence of a hypothesis $H \subseteq \Theta$ over a hypothesis $H' \subseteq \Theta$ is given by*

$$w_{m,n}(H,H') = pl_{m,n}(H)/pl_{m,n}(H').$$

By theorem 4.9 of Kohlas & Monney [27], for all $\theta \in \Theta$, we have

$$pl_{m,n}(\theta) = c \cdot (\prod_{i=1}^{m} l_x(\theta) \cdot \prod_{i=1}^{n} l_{x'}(\theta)) \tag{2.2.3}$$

where c is a positive constant which does not depend on θ. This implies that the general weight of evidence given in this definition generalizes to any type of hypotheses the weight of evidence for simple hypotheses given in equation (2.2.2) :

$$\frac{pl_{m,n}(\theta)}{pl_{m,n}(\theta')} = \prod_{i=1}^{m} w_x(\theta,\theta') \cdot \prod_{i=1}^{n} w_{x'}(\theta,\theta').$$

If the hint $\mathcal{H}_{m,n}$ happens to be precise, i.e. all its focal sets are one-element subsets of Θ, then using theorem 2 it can easily be proved that

$$w_{m,n}(H, H') = \frac{\sum_{\theta \in H} l_{m,n}(\theta)}{\sum_{\theta \in H'} l_{m,n}(\theta)}.$$

Another special case occurs when the hint $\mathcal{H}_{m,n}$ is *consonant*, i.e. all its focal sets are nested [27]. In this case, using theorem 3.7 of Kohlas & Monney [27], we can easily show that

$$w_{m,n}(H, H') = \frac{\max \{l_{m,n}(\theta) : \theta \in H\}}{\max \{l_{m,n}(\theta) : \theta \in H'\}}.$$

The situation where, in addition to x and x', other values are observed can be treated in a similar way.

2.3 Functional and Distribution Models

We have seen that a generalized functional model allows us to compute the weight of evidence for composite hypotheses. The problem is that in general there exist several GFM having the same associated distribution model. This means that if the initial knowledge is given in the form of a distribution model, then the weight of evidence will depend on the GFM that is chosen to represent the given distribution model.

To illustrate this fact, in this section we are going to consider a particular distribution model and analyze two different GFM representing it. However, it should be mentioned that there is no reason why the initial knowledge should necessarily be given in the form of a distribution model. Any functional model that faithfully reflects the mechanism of the generation of the observed data can serve as an initial model.

2.3.1 The Distribution Model of the Problem

Suppose that an urn contains four balls which are either black or white. We successively and randomly draw a ball from the urn and observe its color before placing it back into the urn. Let ξ denote the variable indicating the color of the ball drawn. Suppose that we draw a total of r balls and that m happen to be white whereas $n = r - m$ happen to be black. Based on these observations, what can be inferred on the correct value of the parameter indicating the number of white balls in the urn ? In particular, what is the weight of the evidence represented by the observations with respect to certain hypotheses ?

Let $\Theta = \{0, 1, 2, 3, 4\}$ denote the set of possible values of the parameter (the unknown number of white balls in the urn) and let $X = \{0, 1\}$ denote the set of possible values of the observable variable ξ, where 1 represents white and 0 represents black. The distribution model for this problem is

$$P_\theta(x) = \begin{cases} \theta/4 & \text{if } x = 1 \\ 1 - \theta/4 & \text{if } x = 0 \end{cases} \tag{2.3.4}$$

for all $\theta \in \Theta$. The classical binomial and Bayesian models can be used to make statistical inferences on Θ. Note that for a Bayesian model a prior distribution on Θ is needed. However, if we want to compute weights of evidence for any kind of hypotheses, the distribution model is not sufficient and, as shown above, generalized functional models provide a solution for this problem. In the following two subsections we are going to present two GFMs having P_θ as their associated distribution model.

2.3.2 A GFM Obtained by Conditional Embedding

The generalized functional model that will be presented in this subsection is obtained by the technique of *conditional embedding* proposed by Smets [46]. This method is also described in Shafer [39], where the logic of the construction process is also presented. We take the random perturbation as a label indicating a particular relation between Θ and X. More precisely, the set of possible values of the random perturbation is $\Omega = \{\varphi_1, \varphi_2, \ldots, \varphi_8\}$, where each φ_i is the indicator function of a certain subset S_i of Θ. In other words, $\varphi_i(\theta) = 1$ if $\theta \in S_i$ and 0 otherwise. By definition, the eight sets $S_1, \ldots, S_8$ are

$$S_1 = \{4\},\ S_2 = \{1,4\},\ S_3 = \{2,4\},\ S_4 = \{3,4\},\ S_5 = \{1,2,4\}$$
$$S_6 = \{1,3,4\},\ S_7 = \{2,3,4\},\ S_8 = \{1,2,3,4\}.$$

The distribution P on Ω is defined as

$$P(\varphi_1) = 3/32,\ P(\varphi_2) = 1/32,\ P(\varphi_3) = 3/32,\ P(\varphi_4) = 9/32,$$
$$P(\varphi_5) = 1/32,\ P(\varphi_6) = 3/32,\ P(\varphi_7) = 9/32,\ P(\varphi_8) = 3/32.$$

The last element of the GFM that is still missing is the function $f : \Theta \times \Omega \to X$. This function is defined by

$$f(\theta, \varphi_i) = \varphi_i(\theta).$$

It can easily be proved that the distribution model associated with this GFM is the distributions P_θ given in equation (2.3.4).

If we observe a white ball ($\xi = 1$) in one draw, then the corresponding hint $\mathcal{H}_1$ on Θ has eight focal sets which are

$$\Gamma_1(\varphi_i) = \{\theta \in \Theta : f(\theta, \varphi_i) = 1\} = \{\theta \in \Theta : \varphi_i(\theta) = 1\} = S_i$$

for all $i = 1, \ldots, 8$. The kernel of $\mathcal{H}_1$ is then

$$\mathcal{K} = \{\emptyset, S_1, \ldots, S_8\}$$

and we assume that this ordering of the elements of $\mathcal{K}$ is fixed. According to definition 12, the corresponding incidence matrix is

$$M = \begin{pmatrix} 1&1&1&1&1&1&1&1&1 \\ 0&1&1&1&1&1&1&1&1 \\ 0&0&1&0&0&1&1&0&1 \\ 0&0&0&1&0&1&0&1&1 \\ 0&0&0&0&1&0&1&1&1 \\ 0&0&0&0&0&1&0&0&1 \\ 0&0&0&0&0&0&1&0&1 \\ 0&0&0&0&0&0&0&1&1 \\ 0&0&0&0&0&0&0&0&1 \end{pmatrix}.$$

Of course, the m-value of S_i with respect to $\mathcal{H}_1$ is $P(\varphi_i)$ and hence the restriction of the m-function of $\mathcal{H}_1$ to its kernel is

$$\mu^1 = (0, 3/32, 1/32, 3/32, 9/32, 1/32, 3/32, 9/32, 3/32)'.$$

The hint $\mathcal{H}_1$ is clearly closed and the restriction of its commonality function to its kernel is

$$q^1 = (1, 1, 1/4, 1/2, 3/4, 1/8, 6/32, 12/32, 3/32)'.$$

Now suppose that we observe only white balls in the r draws (i.e. $m = r$ and $n = 0$). Then we have the following result.

Theorem 19 *If μ^m denotes the m-function of the hint*

$$\mathcal{H}^m = \oplus \{\mathcal{H}_1 : i = 1, \ldots, m\},$$

then

$$\begin{aligned}
\mu^m(S_1) &= 1 - (3/32)^m + (3/16)^m + (3/8)^m - (3/4)^m - 2^{-m} - 4^{-m} + 8^{-m} \\
\mu^m(S_2) &= (3/32)^m - (3/16)^m + 4^{-m} - 8^{-m} \\
\mu^m(S_3) &= (3/32)^m - (3/8)^m + 2^{-m} - 8^{-m} \\
\mu^m(S_4) &= (3/32)^m - (3/16)^m - (3/8)^m + (3/4)^m \\
\mu^m(S_5) &= -(3/32)^m + 8^{-m} \\
\mu^m(S_6) &= -(3/32)^m + (3/16)^m \\
\mu^m(S_7) &= -(3/32)^m + (3/8)^m \\
\mu^m(S_8) &= (3/32)^m.
\end{aligned}$$

Proof. A direct application of theorem 16 yields the result. □

This theorem implies the following degrees of plausibility after observing m white balls in $r = m$ draws :

$$
\begin{aligned}
pl_{m,0}(\{1,2\}) &= (1/4)^m + (1/2)^m - (1/8)^m \\
pl_{m,0}(\{2,3\}) &= (3/4)^m + (1/2)^m - (3/8)^m \\
pl_{m,0}(\{1,3\}) &= (3/4)^m + (1/4)^m - (3/16)^m
\end{aligned}
$$

and

$$
\begin{aligned}
pl_{m,0}(\{1\}) &= 4^{-m} \\
pl_{m,0}(\{2\}) &= 2^{-m} \\
pl_{m,0}(\{3\}) &= (3/4)^m \\
pl_{m,0}(\{1,2,3\}) &= (3/32)^m - (3/16)^m - (3/8)^m + (3/4)^m \\
&\quad + 2^{-m} + 4^{-m} - 8^{-m}.
\end{aligned}
$$

Remark that the degree of plausibility of any subset $H \subseteq \Theta$ containing 0 is the same as the degree of plausibility of $H - \{0\}$ because $0 \in S_j$ for every focal set S_j. In addition, the plausibility of any subset $H \subseteq \Theta$ containing 4 is 1 because $4 \in S_j \cap H$ for every focal set S_j. These degrees of plausibility can be used to compute the corresponding weights of evidence. The case where only black balls are observed can be treated in a similar way.

Now consider the situation where we draw $r \geq 2$ balls and $m \geq 1$ happen to be white and $n \geq 1$ happen to be black ($r = m + n$). If $\mathcal{H}_{0,n}$ denotes the hint resulting from the observation of the n black balls and $\mathcal{H}_{m,0}$ denotes the hint resulting from the observation of the m white balls, then

$$\mathcal{H}_{m,n} = \mathcal{H}_{m,0} \oplus \mathcal{H}_{0,n}$$

is the hint corresponding to the observation of all black and white balls. Using theorems 3 and 16, the calculations yield the following degrees of plausibility for the hint $\mathcal{H}_{m,n}$:

$$
\begin{aligned}
pl_{m,n}(\{1,2\}) &= K^{-1}((1/4)^m(3/4)^n - (1/8)^m(3/8)^n + (1/2)^r) \\
pl_{m,n}(\{2,3\}) &= K^{-1}((1/4)^n(3/4)^m - (1/8)^n(3/8)^m + (1/2)^r) \\
pl_{m,n}(\{1,3\}) &= K^{-1}((1/4)^n(3/4)^m + (1/4)^m(3/4)^n - (3/16)^r)
\end{aligned}
$$

for some positive constant K. Once again, these values can be used to compute corresponding weights of evidence as a function of m and n.

2.3.3 A GFM Inspired by Dempster's Structures of the First Kind

In this model suppose that the four balls in the urn are numbered from 1 to 4. In addition, if there are θ white balls in the urn and $\theta \geq 1$, it is assumed that the white balls are numbered from 1 to θ and the black balls are numbered from $\theta + 1$ to 4. Of course, if there is no white ball in the urn, then the black balls are numbered from 1 to 4. This is an important assumption that will permit us to easily specify a functional model. It is similar to the condition which characterizes the so-called *structures of the first kind* introduced by Dempster [11, 13]. Since randomly drawing a ball from the urn is equivalent to randomly selecting the number of a ball, let $\Omega = \{1, \ldots, 4\}$ denote the set of possible values of the random perturbation. Of course its distribution is $P(\omega) = 1/4$ for all $\omega \in \Omega$. To complete the specification of the GFM we define the function $f : \Theta \times \Omega \to X$ by

$$f(\theta, \omega) = \begin{cases} 1 & \text{if } \omega \leq \theta \\ 0 & \text{otherwise} \end{cases}$$

because it is certain that a white ball is observed when the randomly selected number ω is less than or equal to θ.

Let us prove that the distribution model associated with this GFM is the family of distributions P_θ given in equation (2.3.4). If $x = 1$, then we have

$$P(\{\omega \in \Omega : f(\theta, \omega) = 1\}) = P(\{\omega : \omega \leq \theta\}) = \theta/4.$$

Similarly, if $x = 0$, then

$$P(\{\omega \in \Omega : f(\theta, \omega) = 0\}) = P(\{\omega : \omega > \theta\}) = 1 - \theta/4.$$

If we observe a white ball in one draw, then the corresponding hint $\mathcal{H}_1$ on Θ has four focal sets which are

$$\Gamma_1(\omega) = \{\theta \in \Theta : f(\theta, \omega) = 1\} = \{\theta \in \Theta : \omega \leq \theta\} = \{\omega, \omega + 1, \ldots, 4\}$$

for all $\omega \in \Omega = \{1, 2, 3, 4\}$. The m-value of each of these focal sets is $1/4$. Note that $\mathcal{H}_1$ is a consonant hint because its focal sets are nested. Of course, this implies that $\mathcal{H}_1$ is a closed hint and its kernel $\mathcal{K}$ is composed of the five sets

$$K_1 = \emptyset,\ K_2 = \{4\},\ K_3 = \{3, 4\},\ K_4 = \{2, 3, 4\},\ K_5 = \{1, 2, 3, 4\}.$$

With this ordering of the elements of $\mathcal{K}$, the corresponding incidence matrix is

$$M = \begin{pmatrix} 1 & 1 & 1 & 1 & 1 \\ 0 & 1 & 1 & 1 & 1 \\ 0 & 0 & 1 & 1 & 1 \\ 0 & 0 & 0 & 1 & 1 \\ 0 & 0 & 0 & 0 & 1 \end{pmatrix}.$$

Now suppose that we observe only white balls in the r draws, i.e $m = r$ and $n = 0$, and let $\mathcal{H}_{m,0}$ denote the resulting hint, i.e.

$$\mathcal{H}_{m,0} = \oplus \{\mathcal{H}_1 : i = 1, \ldots, m\}.$$

Theorem 20 *The hint $\mathcal{H}_{m,0}$ is a consonant and closed hint whose kernel is $\mathcal{K}$. If μ^m denotes its m-function, then*

$$\mu^m(\{1,2,3,4\}) = (1/4)^m$$
$$\mu^m(\{2,3,4\}) = (1/2)^m - (1/4)^m$$
$$\mu^m(\{3,4\}) = (3/4)^m - (1/2)^m$$
$$\mu^m(\{4\}) = 1 - (3/4)^m.$$

Proof. Since $\mathcal{H}_1$ is closed, a repeated application of theorem 13 shows that the kernel of $\mathcal{H}_{m,0}$ is also $\mathcal{K}$, which also shows that $\mathcal{H}_{m,0}$ is closed and consonant.

If q^1 denotes the restriction to $\mathcal{K}$ of the commonality function of $\mathcal{H}_1$, then

$$q^1 = (1, 1, 3/4, 1/2, 1/4)'.$$

Therefore, using the matrix

$$Q = \begin{pmatrix} 1 & 0 & 0 & 0 & 0 \\ 0 & 1 & 0 & 0 & 0 \\ 0 & 0 & 3/4 & 0 & 0 \\ 0 & 0 & 0 & 1/2 & 0 \\ 0 & 0 & 0 & 0 & 1/4 \end{pmatrix},$$

theorem 16 implies the values for the m-function μ^m given in the theorem. □

If $pl_{m,0}$ denotes the plausibility function of the consonant hint $\mathcal{H}_{m,0}$, then by theorem 3.7 of Kohlas & Monney [27],

$$pl_{m,0}(H) = \max\,\{pl_{m,0}(\theta) : \theta \in H\}$$

for all $H \subseteq \Theta$. Furthermore, if l_1 denotes the likelihood function with respect to the distribution model after observing one white ball, then

$$pl_{m,0}(\theta) = l_1(\theta)^m = (\theta/4)^m$$

for all $\theta \in \Theta$ because it can be shown that $c = 1$ in equation (2.2.3). In particular, we obtain the degrees of plausibility

$$pl_{m,0}(\{1,2\}) = (1/2)^m,\ pl_{m,0}(\{2,3\}) = (3/4)^m,\ pl_{m,0}(\{1,3\}) = (3/4)^m,$$

which in turn can be used to compute weights of evidence. The case where only black balls are observed can be treated in a similar way.

Now consider the situation where we draw $r \geq 2$ balls and $m \geq 1$ happen to be white and $n \geq 1$ happen to be black ($r = m + n$). If $\mathcal{H}_{0,n}$ denotes the hint resulting from the observation of the n black balls, then

$$\mathcal{H}_{m,n} = \mathcal{H}_{m,0} \oplus \mathcal{H}_{0,n}$$

is the hint corresponding to the observation of all black and white balls. Using theorems 3 and 16, the calculations yield the following degrees of plausibility for the hint $\mathcal{H}_{m,n}$:

$$pl_{m,n}(\{1,2\}) = K^{-1}((1/4)^m(3/4)^n + (1/2)^n((1/2)^m - (1/4)^m))$$

$$pl_{m,n}(\{2,3\}) = K^{-1}((1/4)^n(3/4)^m + (1/2)^m((1/2)^n - (1/4)^n))$$

$$pl_{m,n}(\{1,3\}) = K^{-1}((1/4)^m(3/4)^n + (1/4)^n((3/4)^m - (1/4)^m)),$$

where K is a positive constant. These values can in turn can be used to compute corresponding weights of evidence.

Note that the plausibility degrees obtained in this model are different from those obtained in the model considered in the previous subsection. The weights of evidence are also different. In my opinion, the model described in this section is better because it is a more "faithful" representation of the physical process that generated the observed data. This also shows that in general there are several functional models having the same associated distribution model. This means that functional models have a stronger ability to distinguish between concrete situations than distribution models. This is an argument for first finding a "good" functional model describing the situation under investigation, and, if necessary, a distribution model can be derived from it. When possible, functional models are to be preferred to distribution or Bayesian models because they integrate all of the information that is really available.

Finally, let us mention that Smets [49] presents another approach to the problem considered in this section.

2.4 Evidence About a Survival Rate

In this section we are going to generalize the last model based on Dempster's structures of the first kind to the situation where there is an arbitrary number of balls in the urn. However, in order to illustrate the applicability of the theory developed so far, we analyze a real-world example that gives a concrete significance to this urn problem. It is based on a real clinical problem from the early 1980s. The problem was to generate information about the survival rate of newborn babies with a certain critical health problem using a new treatment called extracorporeal membrane oxygenation (Bartlett et al. [1], Ware [51], Begg [2], see also Royall [37]). It had been shown that the survival rate with the old treatment was about 20% and scientists were quite confident that the survival rate using the new treatment could be at least 80%. Based on the new treatment's effect on a few babies, the problem is to gauge the impact of this evidence for a survival rate of more than 80% versus a survival rate of 20%.

Let us analyze this problem with the theory of functional models. Consider the population S of all babies suffering from this specific health problem and let L denote the subset consisting of those babies who would live with the new treatment. Assuming that there are N babies in the entire population S and θ babies in the subpopulation L, the survival rate is defined to be θ/N. Of course the number of surviving babies θ is unknown and it can be any value in the set $\Theta = \{0, 1, 2, \ldots, N\}$. The objective is to draw inferences about θ based on one or several independent observations.

Let ξ denote the variable indicating the result of the new treatment on a baby in S. We define that ξ is 1 when the baby lives and 0 otherwise, so that the set of possible values of ξ is $X = \{0, 1\}$. Also, it is assumed that every baby in the population S has an equal chance of being observed.

Using this information, the following distribution model between Θ and X can be defined : for all $\theta \in \Theta$, we have

$$P_\theta(1) = \theta/N, \quad P_\theta(0) = 1 - (\theta/N). \tag{2.4.5}$$

Given some observations, Royall [37] determines the weight of evidence for the hypothesis $H_2' = \{0.8N\}$ over the hypothesis $H_1' = \{0.2N\}$. We assume that N is large enough for both H_2' and H_1' to represent integer values. Since both of these hypotheses are simple, the distribution model (2.4.5) is sufficient to compute this weight of evidence. However, here we want to determine the weight of evidence for the hypothesis $H_2 = \{\theta : \theta \geq 0.8 \cdot N\}$ over the hypothesis $H_1 = H_1'$. As we have seen, for this kind of hypotheses, a GFM having P_θ as its associated distribution model can be used.

First, it is assumed that the babies in S are numbered from 1 to N in such a way that the babies in L are numbered first. In other words, the babies in L are numbered from 1 to θ and the babies in $S - L$ are numbered from $\theta + 1$ to N. The random perturbation is the number associated with a baby randomly selected from S, so that $\Omega = \{1, \ldots, N\}$ is the set of its possible values. Of course, it is assumed that the baby is randomly selected according to a uniform distribution, i.e. $P(\omega) = 1/N$ for all $\omega \in \Omega$. Now we are in a position to define the function $f : \Theta \times \Omega \to X$ of the GFM by setting

$$f(\theta, \omega) = \begin{cases} 1 & \text{if } \omega \leq \theta \\ 0 & \text{otherwise.} \end{cases}$$

This function gives the result of the new treatment on the baby numbered ω when there are θ babies in the surviving population L. The distribution model associated with this GFM is clearly given by the distributions P_θ given in equations (2.4.5).

Now suppose that we observe a baby who lives under the new treatment. From this evidence, what is the hint $\mathcal{H}_1$ on Θ that can be inferred ? It turns out that $\mathcal{H}_1$ is a consonant hint whose focal sets are

$$\Gamma_1(\omega) = \{\theta \in \Theta : f(\theta, \omega) = 1\} = \{\theta \in \Theta : \omega \leq \theta\} = \{\omega, \omega + 1, \ldots, N\}$$

for all $\omega \in \Omega$. If $[r:s]$ denotes the set of all integers between r and s (the limits are included), then the focal sets of $\mathcal{H}_1$ are the sets $F_i = [i:N]$, $i = 1, \ldots, N$, and the m-value of F_i is $1/N$ for all $i = 1, \ldots, N$. Note that this implies that the hint $\mathcal{H}_1$ is also closed.

Now suppose that the new treatment is given to m babies and that they all live. If $\mathcal{H}_{m,0}$ denotes the hint on Θ induced by these observations, i.e.

$$\mathcal{H}_{m,0} = \oplus \{\mathcal{H}_1 : i = 1, \ldots, m\},$$

then theorem 16 can be used to compute $\mathcal{H}_{m,0}$. However, to add some variety, we are going to perform a direct analysis of this problem. It can be easily noticed that this technique is basically equivalent to the method given by theorem 16.

Since $\mathcal{H}_1$ is closed, the focal sets of $\mathcal{H}_{m,0}$ are again F_i, $i = 1, \ldots, N$, which shows that $\mathcal{H}_{m,0}$ is also consonant.

For $i = 1, \ldots, N$, let $m_i^{(m)}$ denote the m-value of the focal set F_i of $\mathcal{H}_{m,0}$. Then Dempster's rule of combination implies that

$$m_i^{(m)} = (1/N)(i \cdot m_i^{(m-1)} + \sum_{l=1}^{i-1} m_l^{(m-1)}) \tag{2.4.6}$$

for all $i = 1, \ldots, N$. Defining the $N \times N$ matrix

$$T = \begin{pmatrix} 1/N & 0 & \ldots & \ldots & \ldots & \ldots & 0 \\ 1/N & 2/N & 0 & \ldots & \ldots & \ldots & 0 \\ 1/N & 1/N & 3/N & 0 & \ldots & \ldots & \ldots \\ \ldots & \ldots & \ldots & \ldots & \ldots & \ldots & \ldots \\ 1/N & \ldots & 1/N & 1/N & i/N & 0 & \ldots \\ \ldots & \ldots & \ldots & \ldots & \ldots & \ldots & 0 \\ 1/N & 1/N & 1/N & \ldots & \ldots & \ldots & N/N \end{pmatrix}$$

and the vector $m^{(m)} = (m_1^{(m)}, \ldots, m_N^{(m)})'$, the equations (2.4.6) can be written in a more compact way as

$$m^{(m)} = T \cdot m^{(m-1)}. \tag{2.4.7}$$

Note that the matrix T has the remarkable property of being stochastic, i.e. each of its columns sums to 1 and all its entries are non-negative. Considering the focal sets $F_1, \ldots, F_N$ as the N states of a Markov chain, this implies that T is the transition matrix of a Markov chain $Z_t, t = 1, 2, \ldots$ with

$$P(Z_t = F_i) = m_i^{(t)}.$$

Of course, the initial distribution of the chain is

$$P(Z_1 = F_i) = m_i^{(1)} = 1/N$$

for all $i = 1, \ldots, N$. As a consequence of equation (2.4.7) we have

$$m^{(m)} = T^{m-1} \cdot m^{(1)}.$$

Therefore, to compute the m-function of $\mathcal{H}_{m,0}$, the $(m-1)$-th power of the matrix T has to be computed. Fortunately, the matrix T has N different eigenvalues that are all real and by the Jordan decomposition method we can write

$$T = MLM^{-1}$$

where

$$M = \begin{pmatrix} -1 & 0 & \ldots & \ldots & \ldots & \ldots & 0 \\ 1 & -1 & 0 & \ldots & \ldots & \ldots & 0 \\ 0 & 1 & -1 & 0 & \ldots & \ldots & \ldots \\ \ldots & \ldots & \ldots & \ldots & \ldots & \ldots & \ldots \\ 0 & \ldots & 0 & 1 & -1 & 0 & \ldots \\ \ldots & \ldots & \ldots & \ldots & \ldots & \ldots & \ldots \\ 0 & \ldots & \ldots & \ldots & 0 & 1 & -1 \end{pmatrix}$$

and

$$L = \begin{pmatrix} 1/N & 0 & \ldots & \ldots & \ldots & \ldots & 0 \\ 0 & 2/N & 0 & \ldots & \ldots & \ldots & 0 \\ 0 & 0 & 3/N & 0 & \ldots & \ldots & \ldots \\ \ldots & \ldots & \ldots & \ldots & \ldots & \ldots & \ldots \\ 0 & \ldots & \ldots & 0 & i/N & 0 & \ldots \\ \ldots & \ldots & \ldots & \ldots & \ldots & \ldots & \ldots \\ 0 & \ldots & \ldots & \ldots & \ldots & 0 & 1 \end{pmatrix}.$$

This implies that

$$T^{m-1} = M(L^{m-1})M^{-1}$$

and since L^{m-1} is a diagonal matrix whose (i,i)-th element is $(i/N)^{m-1}$ we get

$$m_i^{(m)} = \frac{i^m - (i-1)^m}{N^m}$$

for all $i = 1, \ldots, N$. These values characterize the hint $\mathcal{H}_{m,0}$ coming from the observation of m consecutive successes with the new treatment.

When it is observed that the number of surviving babies m increases and no death is observed, i.e. when m tends to infinity, the m-function of $\mathcal{H}_{m,0}$ tends to the m-function m^* given by $m^*(\{N\}) = 1$ and $m^*(F_i) = 0$ for the other focal sets F_i. But m^* is the m-function of the hint asserting that N is definitely the exact value of the parameter, i.e. every baby will survive with the new treatment. Of course, this result is compatible with our intuition because if we keep observing living babies with the new treatment, then we are more and more inclined to believe that the new treatment will cure all babies. This asymptotic behavior of the hint $\mathcal{H}_{m,0}$ is also a consequence of the fact that the state $\{N\}$ is the only closed class of the Markov chain Z_t, i.e. it is an absorbing state of the chain.

If $pl_{m,0}$ denotes the plausibility function of $\mathcal{H}_{m,0}$, then

$$pl_{m,0}(\theta) = (\theta/N)^m$$

for all $\theta \in \Theta$. As $\mathcal{H}_{m,0}$ is consonant, it is easy to compute the degree of plausibility of any hypothesis $H \subseteq \Theta$ because

$$pl_{m,0}(H) = \max\{pl_{m,0}(\theta) : \theta \in H\}.$$

This in turn allows us to compute the weight of evidence for any pair of hypotheses. For example, for the specific hypotheses H_1 and H_2 defined above, the weight of evidence is

$$w_{m,0}(H_2, H_1) = \frac{1}{0.2^m} = 5^m,$$

which grows exponentially as the number of surviving babies increases.

Now consider the situation where a non-surviving baby is observed. From this observation, what is the hint $\mathcal{H}_0$ on Θ that can be inferred ? It turns out that this is a also a consonant hint whose focal sets are the sets $G_i = [0 : (N - i)]$, $i = 1, \ldots, N$, and the m-value of G_i is $1/N$ for all $i = 1, \ldots, N$. Note that $\mathcal{H}_0$ is also a closed hint.

Now suppose that the new treatment is given to n babies and that none of them survives. If $\mathcal{H}_{0,n}$ denotes the hint on Θ induced by these observations, then the focal sets of $\mathcal{H}_{0,n}$ are again the sets $G_i, i = 1, \ldots, N$ because $\mathcal{H}_0$ is closed, which shows that $\mathcal{H}_{0,n}$ is again a consonant hint. But what is the m-function of this hint ?

If $m_i^{(n)}$ denotes the m-value of the focal set G_i of $\mathcal{H}_{0,n}$, the Jordan decomposition method already applied above can be used to prove that

$$m_i^{(n)} = \frac{i^n - (i-1)^n}{N^n}$$

for all $i = 1, \ldots, N$. Of course, this result can be used to compute $\mathcal{H}_{0,n}$ when n tends to infinity. If $pl_{0,n}$ denotes the plausibility function of $\mathcal{H}_{0,n}$, then

$$pl_{0,n}(\theta) = (1 - (\theta/N))^n$$

for all $\theta \in \Theta$. As $\mathcal{H}_{0,n}$ is consonant, it is easy to compute the degree of plausibility of any hypothesis $H \subseteq \Theta$ because

$$pl_{0,n}(H) = \max\{pl_{0,n}(\theta) : \theta \in H\}.$$

This in turn can be used to compute the weight of evidence for any pair of hypotheses. For example, the weight of evidence for H_2 versus H_1 is

$$w_{0,n}(H_2, H_1) = \frac{0.2^n}{0.8} = 0.25 \cdot 0.2^{n-1},$$

which exponentially decreases with n.

Finally, let us suppose that among $r \geq 2$ babies to which the new treatment is administered, a total of $m \geq 1$ live and a total of $n \geq 1$ do not live $(m + n = r)$. Then

$$\mathcal{H}_{m,n} = \mathcal{H}_{m,0} \oplus \mathcal{H}_{0,n}$$

denotes the hint on Θ induced by these observations. In this case, it turns out that for all (i, j) in

$$\Delta = \{(i,j) \in \Omega \times \Omega : i + j \leq N\}$$

the set $E_{ij} = [i : (N - j)]$ is a focal set of $\mathcal{H}_{m,n}$ and its m-value is

$$m_{m,n}(E_{ij}) = \frac{(i^m - (i-1)^m)(j^n - (j-1)^n)}{K}$$

with

$$K = \sum_{l=1}^{N} (l^n - (l-1)^n)(N-l)^m = \sum_{l=1}^{N} (l^m - (l-1)^m)(N-l)^n.$$

Note that the hint $\mathcal{H}_{m,n}$ is no longer consonant. If $pl_{m,n}$ denotes its plausibility function, then it can be verified that

$$pl_{m,n}(\theta) = \frac{\theta^m (N-\theta)^n}{K}$$

for all $\theta \in \Theta$. In addition, the degree of plausibility of $[0 : r]$ and $[1 : r]$ are the same, and the degree of plausibility of $[r : N]$ and $[r : (N-1)]$ are the same. Furthermore,

$$\begin{aligned} pl_{m,n}([1:r]) &= \frac{\sum_{l=1}^{r} (l^m - (l-1)^m)(N-l)^n}{K} \\ pl_{m,n}([r:N]) &= \frac{\sum_{l=1}^{N-r} (l^n - (l-1)^n)(N-l)^m}{K} \end{aligned}$$

for all r in $\{0, \ldots, N\}$ and

$$pl_{m,n}([r:s]) = pl_{m,n}([1:s]) + pl_{m,n}([r:N]) - 1$$

for all r and s such that $0 \leq r \leq s \leq N$ because, if $sp_{m,n}$ denotes the support function of $\mathcal{H}_{m,n}$, then

$$\begin{aligned} pl_{m,n}([r:s]) &= 1 - sp_{m,n}([0:r-1] \cup [s+1, N]) \\ &= 1 - sp_{m,n}([0:r-1]) - sp_{m,n}([s+1, N]) \\ &= 1 - (1 - pl_{m,n}([r:N])) - (1 - pl_{m,n}([0:s])) \\ &= pl_{m,n}([r, N]) + pl_{m,n}([0:s]) - 1 \\ &= pl_{m,n}([1:s]) + pl_{m,n}([r:N]) - 1. \end{aligned}$$

These results can be used to compute the weight of evidence for any pair of hypotheses. For example, the weight of evidence for H_2 versus H_1 is

$$w_{m,n}(H_2, H_1) = \frac{\sum_{l=1}^{0.2 \cdot N} (l^n - (l-1)^n)(N-l)^m}{0.2^m \cdot 0.8^n \cdot N^r}.$$

The actual observations resulting from two studies about the effectiveness of the new treatment say that among $r = 40$ babies who received the new treatment a total of $m = 39$ babies survived and only $n = 1$ baby did not survive. In this case,

$$w_{39,1}(H_2, H_1) = \frac{\sum_{l=1}^{0.2 \cdot N} (N-l)^{39}}{0.2^{39} \cdot 0.8 \cdot N^{40}},$$

which is a function of unknown value N. However, a finer analysis of this function of N reveals that it stays approximately constant for large values of N, i.e.

$$w_{39,1}(H_2, H_1) \approx 5 \cdot 10^{25}$$

if $N \geq 250$. This means that the weight of the available evidence for H_2 versus H_1 can be taken as $w = 5 \cdot 10^{25}$. In order to give an intuitive interpretation to this number, consider an urn containing two balls which can be either black or white. If W denotes the hypothesis that they are both white and M denotes the hypothesis that there is a mixture of colors, i.e there is one black and one white, then the weight of evidence $w = 5 \cdot 10^{25}$ is approximately the same as the weight of evidence for W versus M when we consecutively observe 85 white balls in 85 random draws from the urn. This is strong evidence for the hypothesis that the survival rate of the new treatment is at least 80% versus the hypothesis that it is 20%.

2.5 Degrees of Support as Weights of Evidence

The most important problem with the use of the traditional likelihood ratio and its extension to plausibility ratio as a measure of the weight of the evidence supporting one hypothesis versus another is that they are not normalized, i.e. the value of these ratios can grow beyond any limit. Furthermore, why should a hypothesis always be compared to another ? It seems more natural to evaluate a hypothesis with respect to the available evidence only, and not with respect to the available evidence and another hypothesis. In other words, the weight of evidence of a hypothesis should be an absolute concept, and not a relative concept, depending only on the available evidence. Therefore, given a hint representing the available evidence, we propose to use the degree of support of a hypothesis as the absolute measure of the weight of the evidence that speaks in favor of the hypothesis. This measure is clearly normalized as a degree of support is a number between 0 and 1. This definition for a weight of evidence is also advocated in Kohlas & Monney [27].

To illustrate these ideas, let us consider the following fictitious example. The internal mechanics of a rudimentary slot machine are partly made of an array of four places marked from 1 to 4, each of which may contain a dime. So the number of dimes in the machine can be anywhere between 0 and 4. The game is played as follows. First a player inserts a nickel in the machine. Then the machine randomly selects one of the four places and returns its contents to the payer, which is either a dime or nothing. Each place has an equal chance of being chosen. One day a player observes the maintenance worker leaving the room after servicing the machine. He asks him how many dimes there are now in the machine and he replies that he cannot reveal this information. However, he does say that he has first removed all the dimes from the machine and then he has consecutively placed a certain number of dimes in the machine, which may be zero, starting from the place number 1 onward, so that there is no empty place between two places containing a dime. For example, if he has put 3 dimes, then there is a dime in place 1, place 2 and place 3. Then the player decides to play and inserts one nickel in the machine, which results in his winning of a dime. The question is: what was the dimes content of the machine just before he played ?

If θ denotes this unknown number of dimes, then the probability of winning a dime is obviously $\theta/4$ and hence

$$l(\theta) = \theta/4, \qquad \theta = 0, 1, \ldots, 4$$

is the likelihood function of θ. Note that the point estimation of θ obtained by the maximum likelihood principle is $\hat{\theta} = 4$, i.e. the machine is completely full of dimes. The following weights of evidence can then be computed :

$$w(1,2) = 1/2, \qquad w(3,4) = 3/4.$$

On the other hand, if we consider the most natural functional model for this situation, i.e. the functional model based on Dempster's structures of the first kind that is equivalent to the one analyzed in subsection 2.3.3, then the information on the number of dimes present in the machine is represented by the hint $\mathcal{H} = (\Omega, P, \Gamma, \Theta)$ where

- $\Theta = \{0, 1, 2, 3, 4\}$ is the set of possible values for the unknown number of dimes in the machine, i.e for the unknown parameter θ
- $\Omega = \{1, 2, 3, 4\}$ is the set representing the 4 possible places for the dimes in the machine
- $P(\omega) = 1/4$ for all $\omega \in \Omega$ because each place has the same chance of being selected
- the mapping

  $$\Gamma : \Omega \longrightarrow \mathcal{P}(\Theta)$$

  is given by

$$\Gamma(1) = \{1, 2, 3, 4\}, \ \Gamma(2) = \{2, 3, 4\}, \ \Gamma(3) = \{3, 4\}, \ \Gamma(4) = \{4\}.$$

From the hint $\mathcal{H}$, it can be seen that every simple hypothesis of the form $\{\theta\}$ with $\theta \neq 4$ has a degree of support equal to zero, which is reasonable because, for these hypotheses, no scenario in the random selection of the place allows us to conclude that a particular specific number of dimes smaller than 4 was present in the machine. Remark that the degree of support of $\theta = 4$ is $1/4$ because if the machine randomly selected the place number 4, then the machine definitely contains 4 dimes. Furthermore, if we consider the hypothesis that at least 1 dime was present, then the degree of support of this hypothesis is 1, which is again very intuitive because we know for sure that this hypothesis is true. If *sp* denotes the support function of $\mathcal{H}$, then $sp(\{2,3,4\}) = 3/4$, which is intuitively explained by the fact that if one of the places 2, 3 or 4 was selected, then there was certainly at least 2 coins in the machine, i.e. the hypothesis $H = \{2,3,4\}$ would be true. In other words, the event $\omega \in \{2,3,4\}$, denoted by E, logically implies that $\theta \in \{2,3,4\}$. Since the probability that E occurs is $3/4$ and E implies H, it is very natural to say that the observation supports the hypothesis H to the degree that the event E occurs, i.e. $3/4$.

In this example, we clearly see the idea that a degree of support is expressing the probability that an argument logically implying the hypothesis is valid. Once again, it is our opinion that this is a proper notion of weight of evidence for a hypothesis.

Note that an observation may also constitute evidence against some hypotheses. For example, winning a dime speaks completely against the hypothesis $\theta = 0$ because the complementary hypothesis $\theta \geq 1$ is certain in view of the observation.

This little example shows that the support function associated with the hint on Θ inferred from an observation gives a clear meaning to the notion of weight of evidence. More generally, the hint inferred on the parameter space can be used to evaluate hypotheses according to their degree of support, a value which is also interpreted as a weight of evidence.

3. Hints on Continuous Frames and Gaussian Linear Systems

All the hints considered so far were defined on a finite frame of discernment Θ. However, in the rest of this book, models where the unknown parameter is continuous will be considered. More precisely, the result of the inference is a hint on $\mathbb{R}^n$. Therefore, we start by presenting the particular features of a hint defined on the continuous frame of discernment $\mathbb{R}^n$.

3.1 Continuous Hints

The type of hints we are going to use is described in the following definition. It has all the characteristics of a hint on a finite frame, the only difference being that they are adapted to the continuous case. The formal theory of continuous hints is presented here for the first time, but Kohlas & Monney [26] contains more information on hints on the frame $\mathbb{R}^n$.

Definition 19 *Let p and n be two integers such that $0 \leq p \leq n$. A continuous hint is a structure of the form*

$$\mathcal{H} = (\mathbb{R}^p, f, \Gamma, \mathbb{R}^n)$$

where

1. *if $p \geq 1$, f is a multivariate probability density function in $\mathbb{R}^p$ and Γ is a mapping from $\mathbb{R}^p$ to the powerset $\mathcal{P}(\mathbb{R}^n)$, i.e.*

$$\Gamma : \mathbb{R}^p \longrightarrow \mathcal{P}(\mathbb{R}^n),$$

such that $\Gamma(\omega) \neq \emptyset$ for all $\omega \in \mathbb{R}^p$, $\Gamma(\omega) \cap \Gamma(\omega') = \emptyset$ whenever $\omega \neq \omega'$ and

$$\cup \{\Gamma(\omega) : \omega \in \mathbb{R}^p\} = \mathbb{R}^n.$$

2. *if $p = 0$, then $\mathbb{R}^0$ is defined to be the real number 0 and f is the Dirac measure at 0. Furthermore, the mapping*

$$\Gamma : \{0\} \longrightarrow \mathcal{P}(\mathbb{R}^n)$$

is given by $\Gamma(0) = \mathbb{R}^n$.

The set $\Gamma(\omega)$ is called the focal set of ω and $\mathbb{R}^n$ is called the frame of the hint. The hint is called precise if $n \geq 1$ and all its focal sets are singletons. The hint is called vacuous if $p = 0$.

Note that f plays the role of the probability measure P for hints on finite frames and that the focal sets $\Gamma(\omega)$ form a partition $\mathcal{Q}$ of $\mathbb{R}^n$.

Definition 20 *Let $\mathcal{H} = (\mathbb{R}^p, f, \Gamma, \mathbb{R}^n)$ be a continuous hint. Then the mapping*

$$m : \mathcal{Q} \longrightarrow [0, +\infty)$$

given by

$$m(F) := f(\Gamma^{-1}(F))$$

is called the basic probability assignment or m-function of $\mathcal{H}$.

Clearly, this definition is similar to the definition of the m-function in the case of hints on finite frames. Note that

$$\int_{\mathbb{R}^p} m(\Gamma(\omega))d\omega = \int_{\mathbb{R}^p} f(\omega)d\omega = 1,$$

which corresponds to the usual property of basic probability assignments on finite frames given by

$$\sum \{m(A) : A \subseteq \Theta\} = 1.$$

As with hints on finite frames of discernment, a support function and a plausibility function are also associated with a continuous hint. If we define

$$u(H) = \{\omega \in \mathbb{R}^p : \Gamma(\omega) \subseteq H\},$$

then the *support function* is the mapping

$$sp : \mathcal{P}(\mathbb{R}^n) \longrightarrow [0, 1]$$

given by

$$sp(H) = P_*(u(H))$$

where P_* is the inner probability measure associated with the density function f. In other words, if $\mathcal{B}$ denotes the Borel σ-algebra in $\mathbb{R}^p$ and P denotes the probability measure associated with the probability density function f, then

$$\begin{aligned} sp(H) &= \sup \{P(K) : K \subseteq u(H), K \in \mathcal{B}\} \\ &= \sup \{\int_K f(\omega)d\omega : K \subseteq u(H), K \in \mathcal{B}\}. \end{aligned}$$

The reason for taking the inner probability measure P_* is that every Borel-measurable subset $K \subseteq u(H)$ supports H and so the degree of support of H is at least $P(K)$ for all such subsets K. Furthermore, if we define

$$v(H) = \{\omega \in \mathbb{R}^p : \Gamma(\omega) \cap H \neq \emptyset\},$$

then the *plausibility function* is the mapping

$$pl : \mathcal{P}(\mathbb{R}^n) \longrightarrow [0, 1]$$

given by

$$pl(H) = P^*(v(H))$$

where P^* is the outer probability measure associated with the probability density function f, i.e.

$$\begin{aligned} pl(H) &= \inf \{P(K) : K \supseteq v(H), K \in \mathcal{B}\} \\ &= \inf \{\int_K f(\omega)d\omega : K \supseteq v(H), K \in \mathcal{B}\}. \end{aligned}$$

The reason for using the outer probability measure P^* is that the relation

$$pl(H) = 1 - sp(H^c)$$

holds for every, non necessarily Borel-measurable, subset $H \subseteq \mathbb{R}^n$. This relation corresponds to the second property of support and plausibility functions mentioned in theorem 1 for hints on finite frames of discernment.

Since $m = f \circ \Gamma^{-1}$ and Γ is injective, it follows that $f = m \circ \Gamma$ and hence

$$sp(H) = \sup \{\int_K m(\Gamma(\omega))d\omega : K \subseteq u(H), K \in \mathcal{B}\}$$

and

$$pl(H) = \inf \{\int_K m(\Gamma(\omega))d\omega : K \supseteq v(H), K \in \mathcal{B}\}.$$

In particular, if both $u(H)$ and $v(H)$ are Borel-measurable, then

$$sp(H) = \int_{u(H)} m(\Gamma(\omega))d\omega \tag{3.1.1}$$

and

$$pl(H) = \int_{v(H)} m(\Gamma(\omega))d\omega. \tag{3.1.2}$$

This shows that the m-function of a continuous hint can be used to compute the support function and the plausibility function of the hint. These functions are completely determined by the m-function of the hint, as was also the case for hints on finite frames of discernment. Furthermore, since these functions can be determined by computing certain integrals of the m-function over some domains, the following definition is quite natural because an absolutely continuous probability measure is also obtained by integrating its density function.

Definition 21 *The m-function of a continuous hint is also called the density of the hint.*

3.2 Gaussian Hints

In the family of continuous hints, there is a special class of hints called Gaussian hints.

Remark 1 *From now on, the multivariate normal probability measure with mean μ and covariance matrix Σ is denoted by $P_{(\mu,\Sigma)}$ and its probability density function is denoted by $f_{\mu,\Sigma}$. It is also assumed that the symmetric covariance matrix Σ is always positive definite.*

Definition 22 *Let p and n be two integers such that $0 \leq p \leq n$. A Gaussian hint on $\mathbb{R}^n$ is a continuous hint of the form*

$$\mathcal{H} = (\mathbb{R}^p, f_{0,\Sigma}, \Gamma, \mathbb{R}^n)$$

where

1. *if $p \geq 1$, the mapping*
$$\Gamma : \mathbb{R}^p \longrightarrow \mathcal{P}(\mathbb{R}^n)$$
is given by
$$\Gamma(\omega) = \{x \in \mathbb{R}^n : Ax = z + \omega\}$$
for some fixed vector z in $\mathbb{R}^p$ and some fixed matrix A in $\mathbb{R}(p, n)$ having full row rank p (both z and A do not depend on ω).
2. *if $p = 0$, then $\mathbb{R}^0$ is defined to be the real number 0 and Σ represents the Dirac measure at 0. Furthermore, the mapping*
$$\Gamma : \{0\} \longrightarrow \mathcal{P}(\mathbb{R}^n)$$
is given by $\Gamma(0) = \mathbb{R}^n$. In other words, $\mathcal{H}$ is the vacuous hint in $\mathbb{R}^n$.

Remark 2 *In a Gaussian hint, the mean of the multivariate normal probability density function $f_{0,\Sigma}$ is always 0. So this density function is completely characterized by the covariance matrix Σ. Therefore, from now on, a Gaussian hint will simply be denoted by*

$$\mathcal{H} = (\mathbb{R}^p, \Sigma, \Gamma, \mathbb{R}^n).$$

The partition $\mathcal{Q}$ of $\mathbb{R}^n$ induced by a Gaussian hint contains the focal sets $\Gamma(\omega)$ which are *parallel linear manifolds* of dimension $n-p$ in $\mathbb{R}^n$. Note that, when $p = 0$, the unique linear manifold of the partition is $\mathbb{R}^n$ itself. Also, the structure of a Gaussian hint is closely related to the concept of normal belief function considered by Dempster [16] and Liu [31]. See also Kohlas & Monney [27] for the definition of a very general notion of hint.

Definition 23 *Let*

$$\mathcal{H}_1 = (\mathbb{R}^p, \Sigma_1, \Gamma_1, \mathbb{R}^n)$$

be a Gaussian hint with

$$\Gamma_1(\omega_1) = \{x \in \mathbb{R}^n : A_1 x = z_1 + \omega_1\}$$

and let

$$\mathcal{H}_2 = (\mathbb{R}^p, \Sigma_2, \Gamma_2, \mathbb{R}^n)$$

be a Gaussian hint with

$$\Gamma_2(\omega_2) = \{x \in \mathbb{R}^n : A_2 x = z_2 + \omega_2\}.$$

Then $\mathcal{H}_1$ and $\mathcal{H}_2$ are equivalent if there exist a regular matrix $T \in \mathbb{R}(p,p)$ such that

1. $A_2 = TA_1$
2. $z_2 = Tz_1$
3. $\Sigma_2 = T\Sigma_1 T'$.

This is a very natural definition because it means that $\mathcal{H}_2$ is simply the hint $\mathcal{H}_1$ expressed with respect to the basis T^{-1} of $\mathbb{R}^p$ instead of the canonical basis. In other words, the perturbation ω_1 in canonical coordinates corresponds to the perturbation $\omega_2 = T\omega_1$ with respect to the basis T^{-1} of $\mathbb{R}^p$. The next theorem says that the corresponding focal sets are the same.

Theorem 21 *If $\mathcal{H}_1$ and $\mathcal{H}_2$ are two equivalent hints and the equivalence is established through the regular matrix T, then $\Gamma_2(T\omega_1) = \Gamma_1(\omega_1)$.*

Proof.

$$\begin{aligned}\Gamma_2(T\omega_1) &= \{x : A_2 x = z_2 + T\omega_1\} \\ &= \{x : TA_1 x = Tz_1 + T\omega_1\} \\ &= \{x : A_1 x = z_1 + \omega_1\} \\ &= \Gamma_1(\omega_1).\end{aligned}$$

□

Theorem 22 *The notion of equivalent Gaussian hints is an equivalence relation in the family of all Gaussian hints on $\mathbb{R}^n$.*

Proof. Using the identity matrix as the regular matrix T implies reflexivity. If $\mathcal{H}_2$ is equivalent to $\mathcal{H}_1$ through the matrix T, then $\mathcal{H}_1$ is equivalent to $\mathcal{H}_2$ through the matrix T^{-1}, which implies symmetry. Finally, if $\mathcal{H}_1$ is equivalent to $\mathcal{H}_2$ through T and $\mathcal{H}_2$ is equivalent to $\mathcal{H}_3$ through Q, then $\mathcal{H}_1$ is equivalent to $\mathcal{H}_3$ through QT, which proves transitivity. □

To keep the notation as simple as possible, we also write $\mathcal{H}_1 = \mathcal{H}_2$ when $\mathcal{H}_1$ and $\mathcal{H}_2$ are equivalent but not necessarily equal.

It is naturally expected that two equivalent Gaussian hints generate the same support and plausibility functions. This property is precisely established in the next theorem.

Theorem 23 *Let*

$$\mathcal{H}_1 = (\mathbb{R}^p, \Sigma_1, \Gamma_1, \mathbb{R}^n)$$

and

$$\mathcal{H}_2 = (\mathbb{R}^p, \Sigma_2, \Gamma_2, \mathbb{R}^n)$$

be two equivalent Gaussian hints. Then $\mathcal{H}_1$ and $\mathcal{H}_2$ generate the same support functions and the same plausibility functions.

Proof. Let sp_1 and sp_2 denote the support functions of $\mathcal{H}_1$ and $\mathcal{H}_2$ and let pl_1 and pl_2 denote their respective plausibility functions. Let T denote the matrix establishing the equivalence of $\mathcal{H}_1$ and $\mathcal{H}_2$. Since $\Gamma_1 = \Gamma_2 \circ T$ by theorem 21, we can easily show that

$$u_2(H) = T(u_1(H)).$$

Let $\tilde{\omega}_2$ be a random variable such that

$$\tilde{\omega}_2 \sim N(0, T\Sigma_1 T')$$

and define the random variable

$$\tilde{\omega}_1 = T^{-1}\tilde{\omega}_2 \sim N(0, \Sigma_1).$$

Then

$$\begin{aligned} sp_2(H) &= \sup \{\int_K f_{0,T\Sigma_1 T'}(\omega_2) d\omega_2 : K \subseteq u_2(H), K \in \mathcal{B}\} \\ &= \sup \{P(\tilde{\omega}_2 \in K) : K \subseteq T(u_1(H)), K \in \mathcal{B}\} \\ &= \sup \{P(T^{-1}\tilde{\omega}_2 \in T^{-1}(K)) : T^{-1}(K) \subseteq u_1(H), K \in \mathcal{B}\} \\ &= \sup \{P(\tilde{\omega}_1 \in T^{-1}(K)) : T^{-1}(K) \subseteq u_1(H), K \in \mathcal{B}\} \\ &= \sup \{P(\tilde{\omega}_1 \in L) : L \subseteq u_1(H), L \in \mathcal{B}\} \\ &= \sup \{\int_L f_{0,\Sigma_1}(\omega_1) d\omega_1 : L \subseteq u_1(H), L \in \mathcal{B}\} \\ &= sp_1(H). \end{aligned}$$

It can be similarly shown that $pl_2(H) = pl_1(H)$, which concludes the proof of the theorem. □

3.3 Precise Gaussian Hints

It is clear that a Gaussian hint on $\mathbb{R}^n$ is precise when the rank of A is n, i.e. when A is a regular matrix of order n. In this case $p = n$ and

$$\Gamma(\omega) = \{A^{-1}z + A^{-1}\omega\}$$

and the partition $\mathcal{Q}$ is simply composed of all points in $\mathbb{R}^n$. These points can be considered as parallel linear manifolds of dimension zero in $\mathbb{R}^n$.

Theorem 24 *Let $\mathcal{H} = (\mathbb{R}^n, \Sigma, \Gamma, \mathbb{R}^n)$ be a precise Gaussian hint where*

$$\Gamma(\omega) = \{x \in \mathbb{R}^n : Ax = z + \omega\}.$$

Then the density of $\mathcal{H}$ is the mapping

$$m : \mathbb{R}^n \longrightarrow \mathbb{R}$$

given by

$$m(x) = f_{0,\Sigma}(Ax - z) = f_{A^{-1}z, A^{-1}\Sigma(A')^{-1}}(x) \tag{3.3.3}$$

Proof. Let us start with the proof of the first equality in (3.3.3). By definition, we have

$$m(x) = f_{0,\Sigma}(\Gamma^{-1}(x)).$$

Since

$$\Gamma(\omega) = \{A^{-1}z + A^{-1}\omega\}$$

it follows that

$$\Gamma^{-1}(x) = Ax - z$$

and hence

$$m(x) = f_{0,\Sigma}(Ax - z). \tag{3.3.4}$$

The second equality in (3.3.3) is a direct consequence of (3.3.4) and proposition C.3 page 255 of Lauritzen [30]. □

The second equality in (3.3.3) shows that the density of a precise Gaussian hint is a normal probability density function with a certain mean and a certain covariance matrix. More precisely, if the precise Gaussian hint is

$$\mathcal{H} = (\mathbb{R}^n, \Sigma, \Gamma, \mathbb{R}^n)$$

with

$$\Gamma(\omega) = \{x \in \mathbb{R}^n : Ax = z + \omega\},$$

then its density is $f_{\mu,V}$ with

$$\mu = A^{-1}z \qquad \text{and} \qquad V = A^{-1}\Sigma(A')^{-1}. \tag{3.3.5}$$

So the density of a precise Gaussian hint is characterized by the two values μ and V, which are respectively called *the mean* and *the covariance* of the precise Gaussian hint. Note that V is also positive definite. So, if we say that $\mathcal{H}$ is a precise Gaussian hint with mean μ and covariance V, this means that there exist A, z, Σ such that

$$\mathcal{H} = (\mathbb{R}^n, \Sigma, \Gamma, \mathbb{R}^n)$$

with

$$\Gamma(\omega) = \{x \in \mathbb{R}^n : Ax = z + \omega\}$$

and the density of $\mathcal{H}$ is $f_{\mu,V}$ with

$$\mu = A^{-1}z \qquad V = A^{-1}\Sigma(A')^{-1}.$$

Now three theorems concerning precise Gaussian hints are formulated.

Theorem 25 *Let $\mathcal{H} = (\mathbb{R}^n, \Sigma, \Gamma, \mathbb{R}^n)$ be a precise Gaussian hint with*

$$\Gamma(\omega) = \{x \in \mathbb{R}^n : Ax = z + \omega\}.$$

Then the support function and the plausibility function of $\mathcal{H}$ coincide on the Borel σ-algebra $\mathcal{B}$ in $\mathbb{R}^n$. Furthermore, if $H \in \mathcal{B}$, then

$$sp(H) = \int_{u(H)} f_{0,\Sigma}(\omega)d\omega. \tag{3.3.6}$$

Proof. First equation (3.3.6) is proved. We have

$$\Gamma(\omega) = \{A^{-1}z + A^{-1}\omega\}$$

and therefore, by theorem 24,

$$m(\Gamma(\omega)) = f_{0,\Sigma}(A(A^{-1}z + A^{-1}\omega) - z) = f_{0,\Sigma}(\omega).$$

Furthermore, it can easily be verified that

$$u(H) = \{-z + AH\},$$

which shows that $u(H) \in \mathcal{B}$ since H is in $\mathcal{B}$. Then, by equation (3.1.1),

$$sp(H) = \int_{u(H)} m(\Gamma(\omega))d\omega = \int_{u(H)} f_{0,\Sigma}(\omega)d\omega,$$

which proves equation (3.3.6).

Since $\Gamma(\omega)$ is a singleton,

$$\Gamma(\omega) \subseteq H \qquad \Leftrightarrow \qquad \Gamma(\omega) \cap H \neq \emptyset,$$

which implies that $u(H) = v(H)$ and hence, by equation (3.1.2),

$$pl(H) = \int_{v(H)} m(\Gamma(\omega))d\omega = \int_{u(H)} f_{0,\Sigma}(\omega)d\omega = sp(H),$$

which proves the theorem. □

Theorem 26 *Let $\mathcal{H} = (\mathbb{R}^n, \Sigma, \Gamma, \mathbb{R}^n)$ be a precise Gaussian hint with*

$$\Gamma(\omega) = \{x \in \mathbb{R}^n : Ax = z + \omega\}.$$

Then the hint

$$\mathcal{H}' = (\mathbb{R}^n, A^{-1}\Sigma A^{-1'}, \Gamma', \mathbb{R}^n)$$

with

$$\Gamma'(\omega') = \{x \in \mathbb{R}^n : Ix = A^{-1}z + \omega'\}$$

is equivalent to $\mathcal{H}$.

Proof. The equivalence is established through the regular matrix A^{-1} in $\mathbb{R}(n,n)$. □

Theorem 27 *Let $\mathcal{H}$ be a precise Gaussian hint with mean μ and covariance V. Then, for any hypothesis $H \in \mathcal{B}$,*

$$sp(H) = \int_H f_{\mu,V}(x)dx.$$

Proof. Since $\mathcal{H}$ has mean μ and covariance V, there exist A, z, Σ such that

$$\mathcal{H} = (\mathbb{R}^n, \Sigma, \Gamma, \mathbb{R}^n)$$

with

$$\Gamma(\omega) = \{x \in \mathbb{R}^n : Ax = z + \omega\}$$

and

$$\mu = A^{-1}z \qquad V = A^{-1}\Sigma(A')^{-1}.$$

Let

$$\mathcal{H}' = (\mathbb{R}^n, V, \Gamma', \mathbb{R}^n)$$

be a precise Gaussian hint with

$$\Gamma'(\omega') = \{x \in \mathbb{R}^n : Ix = \mu + \omega'\}.$$

Then, by theorem 26, the hints $\mathcal{H}$ and $\mathcal{H}'$ are equivalent, and hence by theorem 23 they have the same support function.

If sp' denotes the support function of $\mathcal{H}'$, then by theorem 25

$$sp'(H) = \int_{u'(H)} f_{0,V}(\omega')d\omega'$$

where

$$u'(H) = \{\omega' : \Gamma'(\omega') \subseteq H\} = \{-\mu + H\}.$$

If $\tilde{\omega}'$ is a random variable with probability density function $f_{0,V}$, then

$$\int_{u'(H)} f_{0,V}(\omega')d\omega' = P(\tilde{\omega}' \in u'(H)).$$

Since $x = \mu + \omega'$, the random variable

$$\tilde{x} = \mu + \tilde{\omega}'$$

has a probability density function given by $f_{\mu,V}$. Therefore,

$$\begin{aligned} sp'(H) &= \int_{u'(H)} f_{0,V}(\omega')d\omega' = P(\tilde{\omega}' \in u'(H)) = P(\tilde{\omega}' + \mu \in u'(H) + \mu) \\ &= P(\tilde{x} \in u'(H) + \mu) = P(\tilde{x} \in H) = \int_H f_{\mu,V}(x)dx. \end{aligned}$$

The theorem then follows from the fact that $sp(H) = sp'(H)$ since $\mathcal{H}$ and $\mathcal{H}'$ are equivalent. □

3.4 Gaussian Linear Systems

In this section we present a class of models that we are going to analyze according to the principle of assumption-based reasoning. These models are generalized functional models of a special form.

Let $\mathcal{R}^k$ denote the linear vector space composed of k real numbers arranged in a column. Given a basis $\mathcal{B}$ of $\mathcal{R}^k$, the coordinates of a vector $\boldsymbol{x}$ in $\mathcal{R}^k$ with respect to $\mathcal{B}$ is another column of k real numbers, i.e. an element of $\mathbb{R}^k$ that is typically denoted by the non-boldface letter x. It is important to see the conceptual difference between the vector $\boldsymbol{x}$ in $\mathcal{R}^k$ and its vector of coordinates x with respect to the basis $\mathcal{B}$. The reason for making this difference is that a fixed and uniquely specified data generation model is considered, which, in turn, can have many different representations according to the bases that are being used. As we are going to see, some representations are more appropriate than others for performing the analysis of the model.

The problem is to make inferences about an unknown but fixed vector $\boldsymbol{x}$ in the linear space $\mathcal{R}^n$ based on the observation of a vector $\boldsymbol{z}$ in the linear space $\mathcal{R}^m$. It is assumed that the observed vector $\boldsymbol{z}$ is generated as follows. If $\boldsymbol{x}$ denotes the actual parameter, then there exists a known linear mapping

$$\varphi : \mathcal{R}^n \longrightarrow \mathcal{R}^m$$

such that the observed vector $\boldsymbol{z}$ satisfies the equation

$$\boldsymbol{z} = \varphi(\boldsymbol{x}) - \boldsymbol{\omega} \tag{3.4.7}$$

where $\boldsymbol{\omega}$ is a realization of a random perturbation vector $\tilde{\boldsymbol{\omega}}$ in $\mathcal{R}^m$ having a known normal probability distribution. The distribution of the random vector $\tilde{\boldsymbol{\omega}}$ is expressed with respect to the canonical basis $E_m = \{\boldsymbol{e}_1, \ldots, \boldsymbol{e}_m\}$ of $\mathcal{R}^m$ with

$$e_1 = \begin{pmatrix} 1 \\ 0 \\ \cdot \\ \cdot \\ \cdot \\ 0 \end{pmatrix}, \quad \ldots \quad , e_m = \begin{pmatrix} 0 \\ 0 \\ \cdot \\ \cdot \\ \cdot \\ 1 \end{pmatrix}.$$

More precisely, if $\tilde{\omega}$ denotes the random coordinates vector of $\tilde{\boldsymbol{\omega}}$ with respect to the canonical basis E_m, then $\tilde{\omega}$ is normally distributed with mean 0 and positive definite known covariance matrix Σ :

$$\tilde{\omega} \sim N(0, \Sigma).$$

So the random nature of the perturbation vector $\tilde{\boldsymbol{\omega}}$ is expressed by the random canonical coordinates $\tilde{\omega}$.

Definition 24 *A Gaussian linear system (abbreviated GLS) is an equation of the form*

$$z = \varphi(x) - \omega$$

along with a random vector (in canonical coordinates of $\mathcal{R}^m$)

$$\tilde{\omega} \sim N(0, \Sigma).$$

A Gaussian linear system represents a data generating model for the observed vector z. Of course, a Gaussian linear system is a generalized functional model because it specifies a known relation between the observation, the parameter and the perturbation. With respect to a specific basis of $\mathcal{R}^m$, which is not necessarily the canonical basis, and the canonical basis of $\mathcal{R}^n$, a Gaussian linear system is completely specified by the vector of observations z, the matrix A representing the linear mapping φ and a covariance matrix S.

Several different representations of a same Gaussian linear system are obtained by changing the basis of $\mathcal{R}^m$ that is used. This leads to the following natural definition.

Definition 25 *The representation of a Gaussian linear system with respect to a specific basis of $\mathcal{R}^m$ and the canonical basis of $\mathcal{R}^n$ is called a potential, which is specified by a triple $p = (z, A, \Sigma)$. If the basis of $\mathcal{R}^m$ is the canonical basis, then the potential is called canonical.*

Note that, in a potential, the basis of $\mathcal{R}^n$ is always the canonical basis and so, in a canonical potential, the basis of both $\mathcal{R}^m$ and $\mathcal{R}^n$ are canonical bases.

Definition 26 *Two potentials are equivalent if they represent the same Gaussian linear system.*

Of course, two potentials $p_1 = (z_1, A_1, \Sigma_1)$ and $p_2 = (z_2, A_2, \Sigma_2)$ are equivalent if and only if there exist a regular matrix T in $\mathbb{R}(m, m)$ such that

1. $A_2 = TA_1$
2. $z_2 = Tz_1$
3. $\Sigma_2 = T\Sigma_1 T'$.

This can be used to show that the notion of equivalence for potentials is indeed and equivalence relation. If the two potentials p_1 and p_2 are equivalent, then we still write $p_1 = p_2$ even though they might not be equal, but only equivalent. Furthermore, if we speak of a potential and do not specify the basis of $\mathcal{R}^m$ that is used to represent the Gaussian linear system, then it is implicitly assumed that this is the canonical basis, so that p is indeed a canonical potential.

Note that if the matrix A is the identity matrix, then the potentials containing this matrix are essentially the Gaussian potentials defined and used in Cowell et al. [9]. However, by the presence of the more general matrix

A, there are potentials in our sense that are not potentials in the sense of Cowell et al. [9]. Conversely, there are potentials in the sense of Cowell et al. that are not potentials in our sense because discrete variables are also considered in Cowell et al. [9].

We conclude this chapter with a few important definitions.

Definition 27 *Let*

$$\mathbf{z} = \varphi(\mathbf{x}) - \omega, \qquad \tilde{\omega} \sim N(0, \Sigma)$$

be a Gaussian linear system and let

$$p = (z, A, S), \quad \textit{with} \quad A \in \mathbb{R}(m, n)$$

denote one of its potentials. Then both the Gaussian linear system and the potential are said to be of

1. *the first kind if* $rank(\varphi) = rank(A) = m < n$
2. *the second kind if* $rank(\varphi) = rank(A) < m < n$
3. *the third kind if* $rank(\varphi) = rank(A) = n \leq m$
4. *the fourth kind if* $rank(\varphi) = rank(A) < n \leq m$.

Definition 28 *A Gaussian linear system and its potentials* $p = (z, A, S)$ *with* $A \in \mathbb{R}(m, n)$ *are called proper if* $m \leq n$ *and* $rank(A) = m$.

Clearly, all potentials of the first kind and some of the potentials of the third kind (those for which $rank(A) = n = m$) are proper.

Definition 29 *A Gaussian linear system and its potentials are called underdetermined if they are of the first or second kind, and they are called overdetermined if they are of the third or fourth kind.*

Definition 30 *The Gaussian linear system represented by the canonical potential* $p = (0, 0, 0)$*, where* $\Sigma = 0$ *is taken to represent the Dirac measure at* 0 *in* $\mathbb{R}^m$*, is called vacuous and* $p = (0, 0, 0)$ *is called the vacuous potential.*

4. Assumption-Based Reasoning with Classical Regression Models

In this chapter, a special class of Gaussian linear systems will be analyzed. When expressed in canonical coordinates, these systems turn out to be regular linear regression models, with the slight difference that the distribution of the error term is now assumed to be known (this is a general assumption of functional models). This assumption guarantees that the result of the inference is completely known. Of course, the analysis could be carried out in a similar way when the variance of the distribution of the error term is left unspecified. In this case, the result of the analysis will depend on this unspecified variance. The same remark also holds for any generalized functional model when the distribution of the random perturbation is unknown.

As in the previous chapters, the principle of assumption-based reasoning will be used to generate information about the parameter. As it turns out, this information will be given by some precise Gaussian hint about the unknown parameter, which in turn can be used to determine degrees of support and degrees of plausibility for hypotheses about the parameter. It is shown that this particular way of analyzing regression models leads to the same results as those obtained by least squares or maximum likelihood methods. It is interesting to note that our approach utilizing assumption-based reasoning on a particular functional model yields the same results as the most classical and widely used point estimation principles.

4.1 Classical Linear Regression Models as Special Gaussian Linear Systems

First some notation is introduced.

Definition 31 *Let*

$$\varphi : \mathcal{R}^n \longrightarrow \mathcal{R}^m$$

be a linear mapping. Then the null space of φ is denoted by $\mathcal{N}(\varphi)$ and the range space of φ is denoted by $\mathcal{R}(\varphi)$.

Now consider a Gaussian linear system

$$z = \varphi(x) - \omega \tag{4.1.1}$$

with $\tilde{\omega} \sim N(0, \Sigma)$ in canonical coordinates and such that

$$dim(\mathcal{R}(\varphi)) = n \leq m.$$

In other words, this is a Gaussian linear system of the third kind. If $p = (z, A, \Sigma)$ denotes the canonical potential of this Gaussian linear system, then $A \in \mathbb{R}(m, n)$ and $rank(A) = n \leq m$. These conditions are always assumed throughout this chapter. Furthermore, in canonical coordinates, equation (4.1.1) reads

$$z = Ax - \omega, \tag{4.1.2}$$

which corresponds to a classical linear regression model with observation vector z, design matrix A and error term ω.

4.2 The Principle of the Inference

In this section, assumption-based reasoning is applied to the Gaussian linear system described in the previous section. The goal is to generate inferences about the unknown parameter $\boldsymbol{x}$.

As a consequence of equation (4.1.1), if $\boldsymbol{z} + \boldsymbol{\omega}$ is not in the range of φ, then there is no vector $\boldsymbol{x}$ which could possibly have generated the observed vector $\boldsymbol{z}$. But since we have indeed observed $\boldsymbol{z}$, this means that a perturbation vector $\boldsymbol{\omega}$ for which $\boldsymbol{z}+\boldsymbol{\omega}$ is not in the range of φ is impossible. Therefore, only perturbations $\boldsymbol{\omega}$ such that $\boldsymbol{z} + \boldsymbol{\omega}$ is in $\mathcal{R}(\varphi)$ could possibly have generated $\boldsymbol{z}$. Such a perturbation $\boldsymbol{\omega}$ is called *admissible*.

Definition 32 *Given an observation z, a perturbation vector $\boldsymbol{\omega}$ is called admissible if $\boldsymbol{z} + \boldsymbol{\omega}$ is in the range of φ. The set of all admissible vectors is denoted by Ω.*

So

$$\Omega = \{\boldsymbol{\omega} : \boldsymbol{z} + \boldsymbol{\omega} \in \mathcal{R}(\varphi)\} = \{\boldsymbol{\omega} : \boldsymbol{\omega} \in -\boldsymbol{z} + \mathcal{R}(\varphi)\} = -\boldsymbol{z} + \mathcal{R}(\varphi),$$

which shows that Ω is a linear manifold of dimension n in the vector space $\mathcal{R}^m$ because the dimension of $\mathcal{R}(\varphi)$ is n and $n \leq m$. But

$$dim\ (\mathcal{N}(\varphi)) + dim\ (\mathcal{R}(\varphi)) = n$$

and $dim\ (\mathcal{R}(\varphi)) = n$ imply that $\mathcal{N}(\varphi) = \{0\}$ and so φ is injective. Therefore, the mapping

$$\varphi' : \mathcal{R}^n \longrightarrow \mathcal{R}(\varphi)$$

given by $\varphi'(\boldsymbol{x}) = \varphi(\boldsymbol{x})$ is bijective. Given an admissible perturbation $\boldsymbol{\omega} \in \Omega$, this implies that there is a unique vector $\boldsymbol{u} \in \mathcal{R}^n$ such that $\boldsymbol{z} + \boldsymbol{\omega} = \varphi(\boldsymbol{u})$. Therefore, for this $\boldsymbol{\omega}$, the set $\Gamma(\boldsymbol{\omega})$ of all parameter vectors $\boldsymbol{x}$ satisfying the Gaussian linear system is

$$\begin{aligned}\Gamma(\omega) &:= \{\boldsymbol{x} \in \mathcal{R}^n : \boldsymbol{z} = \varphi(\boldsymbol{x}) - \boldsymbol{\omega}\} \\ &= \{\boldsymbol{x} \in \mathcal{R}^n : \boldsymbol{z} = \varphi(\boldsymbol{x}) + \boldsymbol{z} - \varphi(\boldsymbol{u})\} \\ &= \{\boldsymbol{x} \in \mathcal{R}^n : \varphi(\boldsymbol{x}) = \varphi(\boldsymbol{u})\} \\ &= \{\boldsymbol{u}\}.\end{aligned}$$

This shows that there is *exactly one* vector $\boldsymbol{x}$ induced by the admissible perturbation $\boldsymbol{\omega}$. In other words, if the perturbation was known to be $\boldsymbol{\omega} \in \Omega$, then we could definitely conclude that the correct value of the parameter vector $\boldsymbol{x}$ is the unique vector $\boldsymbol{u}$ in $\Gamma(\boldsymbol{\omega})$. But, of course, we do not know which admissible perturbation $\boldsymbol{\omega}$ has generated the observed vector $\boldsymbol{z}$. All we know is that *before* observing $\boldsymbol{z}$, the distribution of the random perturbation $\tilde{\boldsymbol{\omega}}$ in $\mathcal{R}^m$, expressed in canonical coordinates, was $N(0, \Sigma)$. But, as we have seen, the extra piece of knowledge consisting of the actual observation $\boldsymbol{z}$ implies that only admissible perturbations $\boldsymbol{\omega}$ are eligible to have generated $\boldsymbol{z}$. This information is used by conditioning the distribution of $\tilde{\boldsymbol{\omega}}$ in $\mathcal{R}^m$ on the linear manifold Ω of admissible perturbations.

Remark 3 *From now on, non-boldface symbols represent the canonical coordinates of the corresponding boldface symbols. For further reference, this is called the canonical convention.*

In canonical coordinates and using the canonical convention, given an observation $\boldsymbol{z}$, the set of admissible perturbations is

$$\Omega = \{\omega \in \mathbb{R}^m : \exists\, u \in \mathbb{R}^n \text{ such that } \omega = -z + Au\}, \tag{4.2.3}$$

where A is the matrix representation of the linear mapping φ expressed with respect to the canonical bases of $\mathcal{R}^m$ and $\mathcal{R}^n$. Instead of expressing the Gaussian linear system in canonical coordinates, let us express it using a different basis of $\mathcal{R}^m$, while keeping the canonical basis of $\mathcal{R}^n$. Of course, the Gaussian linear system and the set Ω remain the same, only their *representation* will be different since one of the bases has changed. The new basis on $\mathcal{R}^m$ can be freely selected, but let us choose one that will make it easy for us to condition the distribution of $\tilde{\boldsymbol{\omega}}$ in $\mathcal{R}^m$ on the linear manifold Ω. We consider bases of $\mathcal{R}^m$ of the form

$$B = \begin{pmatrix} A & Z \end{pmatrix} \tag{4.2.4}$$

where $Z \in \mathbb{R}(m, m-n)$ is any matrix composed of $m-n$ linearly independent column vectors such that B is composed of m linearly independent column vectors of $\mathcal{R}^m$. Note that the range space $\mathcal{R}(Z)$ is a supplementary linear space of $\mathcal{R}(A)$.

Now, with respect to the canonical basis E_n for $\mathcal{R}^n$ and the basis B for $\mathcal{R}^m$ and using the canonical convention, the linear equation (4.1.2) becomes

$$B^{-1}z = B^{-1}Ax - B^{-1}\omega \tag{4.2.5}$$

according to a well known result from linear algebra. Note that we still have $\tilde{\omega} \sim N(0, \Sigma)$ in canonical coordinates. Furthermore, if the variable ξ is defined by

$$\xi = B^{-1}\omega,$$

then ξ represents the coordinates of ω with respect to the basis B. Then

$$B^{-1}A = \begin{pmatrix} I \\ 0 \end{pmatrix}$$

because, for instance, B^{-1} times the first column of A represents the coordinates of the first column of A with respect to B, which is clearly e_1. Now, let us define $T_1 \in \mathbb{R}(n, m)$ and $T_2 \in \mathbb{R}(m-n, m)$ such that

$$B^{-1} = \begin{pmatrix} T_1 \\ T_2 \end{pmatrix}.$$

Furthermore, define

$$y_1 = T_1 z \qquad \text{and} \qquad y_2 = T_2 z$$

as well as

$$\xi_1 = T_1\omega \qquad \text{and} \qquad \xi_2 = T_2\omega.$$

Then

$$\xi = \begin{pmatrix} \xi_1 \\ \xi_2 \end{pmatrix}$$

and

$$T_1 A = I \qquad T_2 A = 0.$$

With these definitions, equation (4.2.5) reads

$$\begin{pmatrix} T_1 z \\ T_2 z \end{pmatrix} = \begin{pmatrix} I \\ 0 \end{pmatrix} x - \begin{pmatrix} T_1\omega \\ T_2\omega \end{pmatrix},$$

i.e.

$$T_1 z = x - T_1\omega \qquad T_2 z = -T_2\omega,$$

i.e.

$$y_1 = x - \xi_1 \tag{4.2.6}$$
$$y_2 = -\xi_2. \tag{4.2.7}$$

Expressed with respect to the basis B in $\mathcal{R}^m$ and the canonical basis in $\mathcal{R}^n$, the set of admissible perturbations Ω given in equation (4.2.3) is

$$\begin{aligned} \Omega &= \{B^{-1}\omega \in \mathbb{R}^m : \exists\, u \in \mathbb{R}^n \text{ such that } B^{-1}\omega = -B^{-1}z + B^{-1}Au\} \\ &= \{\xi \in \mathbb{R}^m : \exists\, u \in \mathbb{R}^n \text{ such that } \xi_1 = -y_1 + u,\ \xi_2 = -y_2\} \\ &= \{\xi \in \mathbb{R}^m : \xi_2 = -y_2\}. \end{aligned}$$

Therefore, for an element ξ in Ω, i.e. one such that $\xi_2 = -y_2$, the unique value of x in $\mathbb{R}^n$ induced by equation (4.1.2) is

$$x = y_1 + \xi_1. \tag{4.2.8}$$

Since $\xi = B^{-1}\omega$, the $N(0, \Sigma)$ distribution of $\tilde{\omega}$ is transformed into the

$$N(0, B^{-1}\Sigma(B^{-1})')$$

distribution of $\tilde{\xi}$:

$$\tilde{\xi} \sim N(0, B^{-1}\Sigma(B^{-1})').$$

Now recall that the random variable $\tilde{\xi}$ needs to be conditioned on the linear manifold

$$\Omega = \{\xi \in \mathbb{R}^m : \xi_2 = -y_2\}.$$

This conditioning results in a distribution for $\tilde{\xi}_1$ in $\mathbb{R}^n$ that will be used to find the distribution for x in $\mathbb{R}^n$ using equation (4.2.8). This distribution is simply the m-function of the precise Gaussian hint inferred on the unknown parameter x from the Gaussian linear system. Before proceeding with the actual conditioning, let us first introduce some notation :

$$\begin{aligned} B^{-1}\Sigma(B^{-1})' &= \begin{pmatrix} T_1 \\ T_2 \end{pmatrix} \Sigma \left(T_1' \; T_2'\right) \\ &= \begin{pmatrix} T_1\Sigma T_1' & T_1\Sigma T_2' \\ T_2\Sigma T_1' & T_2\Sigma T_2' \end{pmatrix} = \begin{pmatrix} c_{11} & c_{12} \\ c_{21} & c_{22} \end{pmatrix}. \end{aligned}$$

Theorem 28 *The matrix $c_{22} = T_2\Sigma T_2'$ is regular.*

Proof. Since T_2 has full row rank, this is a classical result of linear algebra. It can be found for instance in Lauritzen [30]. □

By Lauritzen [30], the distribution of $\tilde{\xi}_1$ given $\tilde{\xi}_2 = -y_2$ is

$$N_n(c_{12}c_{22}^{-1}(-y_2), c_{11} - c_{12}c_{22}^{-1}c_{21}). \tag{4.2.9}$$

Theorem 29 *Let*

$$B = (A\; Z)$$

be a basis of $\mathcal{R}^m$ and let

$$T = B^{-1} = \begin{pmatrix} T_1 \\ T_2 \end{pmatrix}.$$

Then the density of the precise Gaussian hint inferred on x has mean

$$\mu_T = T_1 z - T_1\Sigma T_2'(T_2\Sigma T_2')^{-1}T_2 z$$

and covariance

$$\Sigma_T = T_1\Sigma T_1' - (T_1\Sigma T_2')(T_2\Sigma T_2')^{-1}T_2\Sigma T_1'.$$

Proof. This is a direct consequence of equations (4.2.8) and (4.2.9). □

4.3 The Result of the Inference

In this section, the explicit result of the inference on x obtained from a Gaussian linear system represented by the canonical potential $p = (z, A, \Sigma)$ will be given. The following lemma will be used several times.

Lemma 30 *Let F be a $k \times p$ matrix of rank k and E be a $(p-k) \times p$ matrix of rank $p-k$ such that $FE' = 0$. Then*

$$E'(EGE')^{-1}E + G^{-1}F'(FG^{-1}F')^{-1}FG^{-1} = G^{-1} \tag{4.3.10}$$

for any positive definite $p \times p$ matrix G.

Proof. It is well known that a positive definite matrix is regular and its inverse is also positive definite. Since the rank of F' is k and the rank of E' is $p-k$, the matrices EGE' and $FG^{-1}F'$ are also positive definite by a classical theorem of linear algebra. This shows that all inverses appearing in (4.3.10) exist.

Now let us define the matrices

$$M = \begin{pmatrix} F \\ EG \end{pmatrix}$$

and

$$N = E'(EGE')^{-1}E + G^{-1}F'(FG^{-1}F')^{-1}FG^{-1} - G^{-1}.$$

In order to prove (4.3.10), we need to prove that $N = 0$, which is achieved by showing that M is regular and that $MN = 0$.

First, it is shown that M is regular. To do that, we show that the columns of M are linearly independent. Note that the columns of E' form a basis of the null space of F.

Suppose that $Mx = 0$. Then $Fx = 0$ and $EGx = 0$ and so $x \in \mathcal{N}(F)$ and so there exists a vector t such that $x = E't$. But then we have $EGE't = 0$ and since EGE' is regular it follows that $t = 0$, which in turn implies that $x = E't = 0$. This shows that M is regular.

Now, let us prove that $MN = 0$. This is the same as proving that $FN = 0$ and $EGN = 0$. But

$$\begin{aligned} FN &= FE'(EGE')^{-1}E + FG^{-1}F'(FG^{-1}F')^{-1}FG^{-1} - FG^{-1} \\ &= FG^{-1} - FG^{-1} = 0 \end{aligned}$$

because $FE' = 0$, and

$$\begin{aligned} EGN &= EGE'(EGE')^{-1}E + EGG^{-1}F'(FG^{-1}F')^{-1}FG^{-1} - EGG^{-1} \\ &= E + EF'(FG^{-1}F')^{-1}FG^{-1} - E \\ &= E - E = 0 \end{aligned}$$

because $EF' = 0$ since $FE' = 0$. □

Now the following fundamental result can be formulated.

Theorem 31 *The density of the precise Gaussian hint inferred on x has mean*

$$m = (A'\Sigma^{-1}A)^{-1}A'\Sigma^{-1}z$$

and covariance

$$V = (A'\Sigma^{-1}A)^{-1}.$$

Proof. Let B and T_2 be as in theorem 29. If we take $F = T_2$, $E = A'$ and $G = \Sigma^{-1}$, then the conditions of lemma 30 are satisfied :

- $F \in \mathbb{R}(k,p)$ of rank k becomes $T_2 \in \mathbb{R}(m-n,m)$ of rank $m-n$, which is correct because the rows of T_2 are linearly independent since B^{-1} is regular. This implies that $k = m-n$ and $p = m$.
- $E \in \mathbb{R}(p-k,p)$ of rank $p-k$ becomes $A' \in \mathbb{R}(n,m)$ of rank n, which is correct and $p-k = m-(m-n) = n$.
- $FE' = 0$ becomes $T_2A = 0$, which is correct.
- G positive definite becomes Σ^{-1} positive definite, which is correct because Σ is positive definite.

So, by lemma 30, we get

$$A(A'\Sigma^{-1}A)^{-1}A' + \Sigma T_2'(T_2\Sigma T_2')^{-1}T_2\Sigma = \Sigma. \tag{4.3.11}$$

But equation (4.3.11) implies that

$$\Sigma T_2'(T_2\Sigma T_2')^{-1}T_2\Sigma = \Sigma - A(A'\Sigma^{-1}A)^{-1}A', \tag{4.3.12}$$

which in turn implies that

$$\Sigma T_2'(T_2\Sigma T_2')^{-1}T_2 = I - A(A'\Sigma^{-1}A)^{-1}A'\Sigma^{-1}.$$

Then, by theorem 29,

$$\begin{aligned}
\mu_T &= T_1z - T_1\Sigma T_2'(T_2\Sigma T_2')^{-1}T_2z \\
&= T_1(I - \Sigma T_2'(T_2\Sigma T_2')^{-1}T_2z \\
&= T_1(I - (I - A(A'\Sigma^{-1}A)^{-1}A'\Sigma^{-1}))z \\
&= T_1A(A'\Sigma^{-1}A)^{-1}A'\Sigma^{-1}z \\
&= (A'\Sigma^{-1}A)^{-1}A'\Sigma^{-1}z
\end{aligned}$$

since $T_1A = I$. Furthermore, by equation (4.3.12) and theorem 29,

$$\begin{aligned}
\Sigma_T &= T_1\Sigma T_1' - T_1\Sigma T_2'(T_2\Sigma T_2')^{-1}T_2\Sigma T_1' \\
&= T_1\Sigma T_1' - T_1(\Sigma - A(A'\Sigma^{-1}A)^{-1}A')T_1' \\
&= T_1\Sigma T_1' - T_1\Sigma T_1' + T_1A(A'\Sigma^{-1}A)^{-1}A'T_1' \\
&= T_1A(A'\Sigma^{-1}A)^{-1}(T_1A)' \\
&= (A'\Sigma^{-1}A)^{-1}
\end{aligned}$$

since $T_1A = I$. This concludes the proof of the theorem. □

This result is very important because it says that the density of the inferred hint about the unknown parameter vector $\boldsymbol{x}$, expressed in terms of the canonical coordinates x, depends only on the available information z, A, Σ of the canonical potential $p = (z, A, \Sigma)$ representing the Gaussian linear system

$$\boldsymbol{z} = \varphi(\boldsymbol{x}) - \omega, \quad \tilde{\omega} \sim N(0, \Sigma).$$

It is very interesting to notice that the density of the hint for x obtained in theorem 31 corresponds exactly to the density of the classical Gauss-Markov estimator, i.e. the least squares estimator. In this book, the mean and the covariance matrix are obtained in a dramatically different way since there is no minimization process involved, in contrast with the theory of least squares estimation. Furthermore, these values also result from the application of the maximum likelihood principle. In particular, theorem 31 implies that the value of x with *maximum density of support* is

$$(A'\Sigma^{-1}A)^{-1}A'\Sigma^{-1}z,$$

thereby showing that the principle of maximum density of support, the maximum likelihood method and the least squares method all yield the same point estimation of x. Note that the density of the precise hint on x can be called a fiducial distribution in view of the method that was used to derive it.

Let us also mention that the location and scale parameter model

$$z = x - \sigma\omega,$$

where z is the observation, x and σ are the unknown parameters and ω is a random perturbation with known probability density function $g(\omega)$, is considered as a functional model by Kohlas & Monney [27]. The analysis of this model is done by means of the assumption-based reasoning principle.

Corollary 32 *If $\Sigma = \sigma^2 I$ for a known positive value σ^2, then the density of the precise Gaussian hint inferred on x has mean*

$$m = (A'A)^{-1}A'z$$

and covariance

$$V = \sigma^2(A'A)^{-1}.$$

Proof. By theorem 31,

$$m = (A'(\sigma^2)^{-1})IA)^{-1}A'(\sigma^2)^{-1}Iz = \sigma^2(A'A)^{-1}(\sigma^2)^{-1}A'z = (A'A)^{-1}A'z$$

and

$$V = (A'(\sigma^2)^{-1}A)^{-1} = \sigma^2(A'A)^{-1},$$

which proves the corollary. □

4.4 Changing Basis

Now let us consider another basis

$$C = (R\ Z)$$

of $\mathcal{R}^m$, where R is a basis of the range of A and Z is a matrix as in equation (4.2.4). In this section, it will be proved that the inferred hint on x is still the one given in theorem 31 if the analysis is performed with respect to the basis C instead of the basis B defined in equation (4.2.4). This is an important result asserting that the inference obtained from a Gaussian linear system

$$z = \varphi(x) - \omega, \qquad \tilde{\omega} \sim N(0, \Sigma) \tag{4.4.13}$$

does not depend on the potential used to represent it, i.e. it is the same for all (equivalent) potentials representing the Gaussian linear system. If we define

$$C^{-1} = \begin{pmatrix} S_1 \\ S_2 \end{pmatrix}$$

and

$$\eta = C^{-1}\omega$$

and

$$\eta_1 = S_1\omega \qquad \eta_2 = S_2\omega,$$

then

$$\eta = \begin{pmatrix} \eta_1 \\ \eta_2 \end{pmatrix}$$

and the Gaussian linear system (4.4.13), when expressed with the canonical basis of $\mathcal{R}^n$ and the basis C of $\mathcal{R}^m$, becomes

$$C^{-1}z = C^{-1}Ax - C^{-1}\omega. \tag{4.4.14}$$

Note that we still have $\tilde{\omega} \sim N(0, \Sigma)$ in canonical coordinates. There exits a unique regular matrix U in $\mathbb{R}(n, n)$ such that

$$C^{-1}A = \begin{pmatrix} S_1 A \\ S_2 A \end{pmatrix} = \begin{pmatrix} U \\ 0 \end{pmatrix}$$

because $C^{-1}A$ is the matrix of the coordinates of the vectors in A with respect to the basis R and R is a basis of $\mathcal{R}(A)$. Therefore $U = S_1A$ is regular and $S_2A = 0$. This implies that equation (4.4.14) becomes

$$\begin{aligned} S_1 z &= Ux - S_1\omega \\ S_2 z &= -S_2\omega. \end{aligned}$$

Defining $w_1 = S_1 z$ and $w_2 = S_2 z$ permits to write

$$w_1 = Ux - \eta_1 \tag{4.4.15}$$
$$w_2 = -\eta_2 \tag{4.4.16}$$

and it can be easily shown that

$$\Omega = \{\eta : \eta_2 = -w_2\}.$$

Since $\eta = C^{-1}\omega$, the $N(0, \Sigma)$ distribution of $\tilde{\omega}$ in canonical coordinates is transformed into

$$\tilde{\eta} \sim N(0, C^{-1}\Sigma(C^{-1})')$$

with respect to the basis C of $\mathcal{R}^m$. Now the random variable $\tilde{\eta}$ needs to be conditioned on the linear manifold

$$\Omega = \{\eta : \eta_2 = -w_2\}.$$

For this purpose, let us once more introduce some notation :

$$C^{-1}\Sigma(C^{-1})' = \begin{pmatrix} S_1 \\ S_2 \end{pmatrix} \Sigma \,(S_1' \; S_2')$$
$$= \begin{pmatrix} S_1\Sigma S_1' & S_1\Sigma S_2' \\ S_2\Sigma S_1' & S_2\Sigma S_2' \end{pmatrix} = \begin{pmatrix} h_{11} & h_{12} \\ h_{21} & h_{22} \end{pmatrix}.$$

As before, the conditional distribution of $\tilde{\eta}_1$ given $\tilde{\eta}_2 = -w_2$ is

$$N_n(h_{12}h_{22}^{-1}(-w_2), h_{11} - h_{12}h_{22}^{-1}h_{21}).$$

Given an element η in Ω, i.e. one such that $\eta_2 = -w_2$, the unique value of x in $\mathbb{R}^n$ induced by the equation (4.4.15) is

$$x = U^{-1}w_1 + U^{-1}\eta_1.$$

This implies that the density of the precise Gaussian hint inferred on x has mean

$$U^{-1}w_1 + U^{-1}h_{12}h_{22}^{-1}(-w_2)$$

and covariance

$$U^{-1}(h_{11} - h_{12}h_{22}^{-1}h_{21})(U^{-1})'.$$

Theorem 33 *Let*

$$C = (R\; Z)$$

be a basis of $\mathcal{R}^m$ such that $\mathcal{R}(R) = \mathcal{R}(A)$ and let

$$S = C^{-1} = \begin{pmatrix} S_1 \\ S_2 \end{pmatrix}.$$

Then the density of the precise Gaussian hint inferred on x has mean

$$\mu_S = (S_1A)^{-1}S_1z - (S_1A)^{-1}S_1\Sigma S_2'(S_2\Sigma S_2')^{-1}S_2z$$

and covariance

$$\Sigma_S = (S_1A)^{-1}(S_1\Sigma S_1' - (S_1\Sigma S_2')(S_2\Sigma S_2')^{-1}S_2\Sigma S_1')((S_1A)^{-1})'.$$

Proof. This is a direct consequence of the developments given in this section. □

Theorem 34 *Using the basis C given in theorem 33, the density of the precise Gaussian hint inferred on x has mean*

$$m = (A'\Sigma^{-1}A)^{-1}A'\Sigma^{-1}z$$

and covariance

$$V = (A'\Sigma^{-1}A)^{-1}.$$

Proof. By applying lemma 30 with $F = S_2$, $E = A'$ and $G = \Sigma^{-1}$, we obtain

$$A(A'\Sigma^{-1}A)^{-1}A' + \Sigma S_2'(S_2\Sigma S_2')^{-1}S_2\Sigma = \Sigma,$$

and hence

$$\Sigma S_2'(S_2\Sigma S_2')^{-1}S_2\Sigma = \Sigma - A(A'\Sigma^{-1}A)^{-1}A'$$

and

$$\Sigma S_2'(S_2\Sigma S_2')^{-1}S_2 = I - A(A'\Sigma^{-1}A)^{-1}A\Sigma^{-1}.$$

Then, according to theorem 33,

$$\begin{aligned}\mu_S &= (S_1A)^{-1}S_1z - (S_1A)^{-1}S_1\Sigma S_2'(S_2\Sigma S_2')S_2z\\ &= (S_1A)^{-1}S_1z - (S_1A)^{-1}S_1(I - A(A'\Sigma^{-1}A)^{-1}A\Sigma^{-1})z\\ &= (S_1A)^{-1}S_1z - (S_1A)^{-1}S_1z + (S_1A)^{-1}(S_1A)(A'\Sigma^{-1}A)^{-1}A\Sigma^{-1}z\\ &= (A'\Sigma^{-1}A)^{-1}A\Sigma^{-1}z,\end{aligned}$$

and

$$\begin{aligned}\Sigma_S &= (S_1A)^{-1}(S_1\Sigma S_1{}' - S_1\Sigma S_2'(S_2\Sigma S_2')^{-1}S_2\Sigma S_1')((S_1A)^{-1})'\\ &= (S_1A)^{-1}(S_1\Sigma S_1' - S_1(\Sigma - A(A'\Sigma^{-1}A)^{-1}A')S_1')((S_1A)^{-1})'\\ &= (S_1A)^{-1}(S_1\Sigma S_1') - S_1\Sigma S_1' + S_1A(A'\Sigma^{-1}A)^{-1}A'S_1')((S_1A)^{-1})'\\ &= (S_1A)^{-1}(S_1A)(A'\Sigma^{-1}A)^{-1}A'S_1'((S_1A)^{-1})'\\ &= (A'\Sigma^{-1}A)^{-1}(S_1A)'((S_1A)')^{-1}\\ &= (A'\Sigma^{-1}A)^{-1},\end{aligned}$$

which concludes the proof of the theorem. □

This last result shows that the inferred hint on x is the same when the analysis is performed with respect to *any* basis

$$K = (R\ Z)$$

of $\mathcal{R}^m$ provided that the range of R is the same as the range of A. This condition on R is necessary because otherwise we do not know how to compute the conditional distribution of the random perturbation on the linear manifold Ω of admissible perturbations. The class $\mathcal{C}$ of all bases of $\mathcal{R}^m$ is composed of two disjoint subclasses : the subclasses $\mathcal{C}_1$ consisting of all bases for which the first n columns form a basis of the range of A, and the subclass $\mathcal{C}_2 = \mathcal{C} - \mathcal{C}_1$ consisting of all bases for which this condition is not satisfied. We have been able to find the result of the inference of the Gaussian linear system when it is expressed with respect to a basis in $\mathcal{C}_1$. However, when it is expressed with respect to a basis in $\mathcal{C}_2$, we simply do not know how to perform the assumption-based reasoning analysis of the Gaussian linear system in a straightforward manner. The point is that some bases are more appropriate than others to perform the analysis. However, for a basis in $\mathcal{C}_2$, we define the result of the inference to be the same as the one obtained from a basis in $\mathcal{C}_1$, which allows us to say that the result of the inference obtained from a Gaussian linear system is independent of the potential used to represent it, i.e. it is the same for all (equivalent) potentials representing the Gaussian linear system.

4.5 Permissible Bases and Admissible Matrices

According to the results of the last section, it is quite natural to give the following definition.

Definition 33 *Given a matrix A in $\mathbb{R}(m,n)$ with $rank(A) = n \leq m$, a basis*

$$K = (R\ Z)$$

in $\mathbb{R}(m,m)$ such that $\mathcal{R}(R) = \mathcal{R}(A)$ is called permissible.

Since Σ is positive definite, the Choleski decomposition of Σ implies that there exists a regular matrix Y such that

$$\Sigma = YY'.$$

Now let us express the original Gaussian linear system (4.1.1) with respect to the basis Y of $\mathcal{R}^m$ and the canonical basis of $\mathcal{R}^n$. We get

$$(Y^{-1}A)x = (Y^{-1}z) + (Y^{-1})\omega \tag{4.5.17}$$

and defining

$$\nu = Y^{-1}\omega$$

permits to write

$$(Y^{-1}A)x = (Y^{-1}z) + \nu$$

with $\tilde{\nu} \sim N(0, I)$ because the covariance matrix of ν is

$$Y^{-1}\Sigma(Y^{-1})' = Y^{-1}YY'(Y')^{-1} = I.$$

This shows that the covariance matrix of the perturbation becomes the identity matrix when the Gaussian linear system (4.1.1) is expressed with respect to the basis Y. It is now very easy to find a permissible basis using the matrix Y. Let $Q \in \mathbb{R}(m, m-n)$ be any matrix composed of $m - n$ linearly independent column vectors such that the matrix

$$M = \begin{pmatrix} Y^{-1}A & Q \end{pmatrix}$$

is regular. Then the basis

$$K = YM$$

is permissible because the first n columns of K are given by the matrix

$$R = Y(Y^{-1}A) = A,$$

and so the range of R is the same as the range of A.

Definition 34 *Given a matrix A in $\mathbb{R}(m,n)$ with $rank(A) = n \leq m$, a matrix T in $\mathbb{R}(m,m)$ is called admissible if T^{-1} exists and is a permissible basis.*

This definition implies that the matrix inversion operator is a bijection between admissible matrices and permissible bases. The following two results give more information on admissible matrices.

Theorem 35 *Let A be a matrix in $\mathbb{R}(m,n)$ with $rank(A) = n \leq m$ and let*

$$T = \begin{pmatrix} T_1 \\ T_2 \end{pmatrix}$$

be a matrix in $\mathbb{R}(m,m)$ with $T_1 \in \mathbb{R}(n,m)$ and $T_2 \in \mathbb{R}(m-n,m)$. If T is admissible, then T_1A is regular and $T_2A = 0$. Conversely, let

$$A = \begin{pmatrix} A_1 \\ A_2 \end{pmatrix}$$

where A_1 denotes the first n rows of A. If T and A_1 are regular and $T_2A = 0$, then T is admissible.

Proof. If T is admissible then T^{-1} is a permissible basis and hence there exist matrices U_1 and U_2 such that

$$T^{-1} = (U_1\, U_2)$$

and $U_1 = AV$ for an invertible matrix V in $\mathbb{R}(n,n)$. Then TA expresses the columns of A in terms of the basis

$$T^{-1} = (U_1\, U_2)\,.$$

Since

$$A = (U_1 V^{-1}) + (U_2 0)$$

it follows that

$$TA = \begin{pmatrix} T_1 A \\ T_2 A \end{pmatrix} = \begin{pmatrix} V^{-1} \\ 0 \end{pmatrix}.$$

This shows that $T_1 A$ is regular and $T_2 A = 0$.

Now suppose that T and A_1 are regular and suppose that $T_2 A = 0$. We must show that T^{-1} is a permissible basis. Let

$$T = \begin{pmatrix} T_{11} & T_{12} \\ T_{21} & T_{22} \end{pmatrix}$$

be a partition of T such that

$$T_1 = (T_{11}\, T_{12})$$

and

$$T_2 = (T_{21}\, T_{22})\,.$$

It is well known that the inverse of T is given by

$$T^{-1} = \begin{pmatrix} T_{11} & T_{12} \\ T_{21} & T_{22} \end{pmatrix}^{-1} = \begin{pmatrix} R_1 & R_3 \\ R_2 & R_4 \end{pmatrix}$$

where

$$R_1 = (T_{11} - T_{12}T_{22}^{-1}T_{21})^{-1}$$

and

$$R_2 = -T_{22}T_{21}(T_{11} - T_{12}T_{22}^{-1}T_{21})^{-1}$$

(see e.g. Lauritzen [30]). Now, $T_2 A = 0$ implies that

$$(T_{21}\, T_{22}) \begin{pmatrix} A_1 \\ A_2 \end{pmatrix} = 0,$$

i.e.

$$T_{21}A_1 + T_{22}A_2 = 0,$$

i.e.

$$T_{22}A_2 = -T_{21}A_1,$$

and so

$$A_2 = -T_{22}^{-1}T_{21}A_1.$$

Then

$$\begin{aligned} R_2 &= -T_{22}T_{21}(T_{11} - T_{12}T_{22}^{-1}T_{21})^{-1} \\ &= (-T_{22}^{-1}T_{21}A_1)(A_1^{-1})(T_{11} - T_{12}T_{22}^{-1}T_{21})^{-1} \\ &= A_2(A_1^{-1}(T_{11} - T_{12}T_{22}^{-1}T_{21})^{-1}) \end{aligned}$$

and

$$R_1 = (T_{11} - T_{12}T_{22}^{-1}T_{21})^{-1} = A_1(A_1^{-1}(T_{11} - T_{12}T_{22}^{-1}T_{21})^{-1}).$$

Therefore, defining

$$L = A_1^{-1}(T_{11} - T_{12}T_{22}^{-1}T_{21})^{-1}$$

permits to write

$$\begin{pmatrix} R_1 \\ R_2 \end{pmatrix} = \begin{pmatrix} A_1 \\ A_2 \end{pmatrix} L = AL,$$

which shows that the first n columns of T^{-1} form a basis of the range of A, which means that T is admissible. □

In theorem 35, note that the condition that A_1 is regular is not essential because we can always rearrange the vectors $e_1, \ldots, e_m$ of the canonical basis E_m of $\mathcal{R}^m$ such that the matrix A representing the linear mapping φ with respect to E_n and E_m is such that A_1 is regular because the dimension of the range of φ is n. If this rearrangement needs to be done, the original covariance matrix Σ also needs to be rearranged accordingly.

Lemma 36 *Let T_2 in $\mathbb{R}(m-n, m)$ be a matrix of rank $m-n$. Then*

$$T_2A = 0 \Leftrightarrow \mathcal{N}(T_2) = \mathcal{R}(A) \Leftrightarrow \mathcal{R}(T_2') = \mathcal{N}(A').$$

Proof. Let us start with the first equivalence. First remark that

$$\begin{aligned} dim(\mathcal{N}(T_2)) &= m - dim(\mathcal{R}(T_2)) \\ &= m - Rank(T_2) \\ &= m - (m-n) \\ &= n. \end{aligned}$$

Now, if $T_2A = 0$, then each column of A is in $\mathcal{N}(T_2)$ and since these n columns are linearly independent $dim(\mathcal{N}(T_2)) = n$ implies that the columns of A form a basis of $\mathcal{N}(T_2)$ and therefore $\mathcal{R}(A) = \mathcal{N}(T_2)$. Conversely, if $\mathcal{R}(A) = \mathcal{N}(T_2)$ then clearly $T_2A = 0$ because each column of A is in $\mathcal{N}(T_2)$.

Now let us prove that $T_2A = 0$ is equivalent to $\mathcal{R}(T_2') = \mathcal{N}(A')$. If $T_2A = 0$ then $A'T_2' = 0$ and so $\mathcal{R}(T_2')$ is a linear subspace of $\mathcal{N}(A')$. But since

$$dim\ \mathcal{N}(A') + dim\ \mathcal{R}(A') = m$$

and $dim\ \mathcal{R}(A') = n$, it follows that $dim\ \mathcal{N}(A') = m - n$. Since the $m - n$ columns of T_2' are linearly independent by hypothesis, it follows that

$$\mathcal{R}(T_2') = \mathcal{N}(A).$$

Conversely, if $\mathcal{R}(T_2') = \mathcal{N}(A')$ then $A'T_2' = 0$ and hence $T_2A = 0$. □

4.6 Different Representations of the Result of the Inference

Let us still consider a Gaussian linear system of the third kind with canonical potential $p = (z, A, \Sigma)$ and $A \in \mathbb{R}(m, n)$ and $rank(A) = n \leq m$. Furthermore, assume without loss of generality (see the previous section) that the matrix A_1 composed of the first n rows of A is regular, i.e.

$$A = \begin{pmatrix} A_1 \\ A_2 \end{pmatrix}$$

where $A_1 \in \mathbb{R}(n, n)$ is regular and $A_2 \in \mathbb{R}(m - n, n)$. In this section, several representations of the mean and covariance of the density of the precise Gaussian hint inferred from the Gaussian linear system will be given. These expressions are obtained by using several different admissible matrices T.

4.6.1 A Representation Derived from Linear Dependencies

Let us define the matrix Λ in $\mathbb{R}(m - n, n)$ by the equation

$$\Lambda = A_2 A_1^{-1}.$$

Since $A_2 = \Lambda A_1$, this means that the matrix Λ is derived from the linear dependencies that prevail among the rows of A.

Theorem 37 *The matrix*

$$T = \begin{pmatrix} I_n & 0 \\ -\Lambda & I_{m-n} \end{pmatrix}$$

is admissible.

Proof. According to theorem 35, it is sufficient to show that T is regular and $T_2 A = 0$ where

$$T_2 = \begin{pmatrix} -\Lambda & I_{m-n} \end{pmatrix}.$$

But

$$\begin{aligned} T_2 A &= \begin{pmatrix} -\Lambda & I_{m-n} \end{pmatrix} \begin{pmatrix} A_1 \\ A_2 \end{pmatrix} \\ &= -\Lambda A_1 + A_2 \\ &= -A_2 A_1^{-1} A_1 + A_2 \\ &= 0 \end{aligned}$$

and it remains to prove that T is regular. But it can easily be verified that

$$T^{-1} = \begin{pmatrix} I_n & 0 \\ \Lambda & I_{m-n} \end{pmatrix}.$$

□

Let z_1 denote the first n elements of z and let z_2 denote the last $m-n$ elements of z, so that

$$z = \begin{pmatrix} z_1 \\ z_2 \end{pmatrix}.$$

Furthermore, let us define

$$\Sigma = \begin{pmatrix} \Sigma_{11} & \Sigma_{12} \\ \Sigma_{21} & \Sigma_{22} \end{pmatrix}$$

where

$$\Sigma_{11} \in \mathbb{R}(n,n) \quad \Sigma_{12} \in \mathbb{R}(n, m-n)$$

and

$$\Sigma_{21} \in \mathbb{R}(m-n, n) \quad \Sigma_{22} \in \mathbb{R}(m-n, m-n).$$

Theorem 38 *The density of the precise Gaussian hint inferred on x has mean*

$$\begin{aligned} m = {} & A_1^{-1} z_1 - (A_1^{-1}(-\Sigma_{11}(A_1')^{-1}A_2' + \Sigma_{12}) \\ & (A_2 A_1^{-1} \Sigma_{11} (A_1')^{-1} A_2' - A_2 A_1^{-1} \Sigma_{12} - \Sigma_{21} (A_1')^{-1} A_2' + \Sigma_{22})^{-1} \\ & (-A_2 A_1^{-1} z_1 + z_2)) \end{aligned}$$

and covariance

$$\begin{aligned} V = {} & A_1^{-1}(\Sigma_{11} - (-\Sigma_{11}(A_1')^{-1}A_2' + \Sigma_{12}) \\ & (A_2 A_1^{-1} \Sigma_{11} (A_1')^{-1} A_2' - A_2 A_1^{-1} \Sigma_{12} - \Sigma_{21} (A_1')^{-1} A_2' + \Sigma_{22})^{-1} \\ & (-A_2 A_1^{-1} \Sigma_{11} + \Sigma_{21}))(A_1^{-1})'. \end{aligned}$$

Proof. Since

$$T = \begin{pmatrix} I_n & 0 \\ -\Lambda & I_{m-n} \end{pmatrix}$$

is an admissible matrix according to theorem 37, it follows that $C = T^{-1}$ is a permissible basis. By theorem 33, defining

$$S = C^{-1} = (T^{-1})^{-1} = T = \begin{pmatrix} I_n & 0 \\ -\Lambda & I_{m-n} \end{pmatrix} = \begin{pmatrix} S_1 \\ S_2 \end{pmatrix},$$

i.e.

$$S_1 = (I_n \; 0)$$

and

$$S_2 = (-\Lambda \; I_{m-n}),$$

it can be concluded that

$$m = (S_1 A)^{-1} S_1 z - (S_1 A)^{-1} S_1 \Sigma S_2' (S_2 \Sigma S_2')^{-1} S_2 z$$

and

$$V = (S_1 A)^{-1} (S_1 \Sigma S_1' - (S_1 \Sigma S_2')(S_2 \Sigma S_2')^{-1} S_2 \Sigma S_1')((S_1 A)^{-1})'.$$

Since

$$S_1 z = (I_n \; 0) \begin{pmatrix} z_1 \\ z_2 \end{pmatrix} = z_1$$

and

$$S_2 z = (-\Lambda \; I_{m-n}) \begin{pmatrix} z_1 \\ z_2 \end{pmatrix} = -\Lambda z_1 + z_2,$$

it follows that

$$m = A_1^{-1} z_1 - (A_1^{-1}(-\Sigma_{11}\Lambda' + \Sigma_{12})(\Lambda\Sigma_{11}\Lambda' - \Lambda\Sigma_{12} - \Sigma_{21}\Lambda' + \Sigma_{22})^{-1} (-\Lambda z_1 + z_2))$$

and

$$V = A_1^{-1}(\Sigma_{11} - (-\Sigma_{11}\Lambda' + \Sigma_{12})(\Lambda\Sigma_{11}\Lambda' - \Lambda\Sigma_{12} - \Sigma_{21}\Lambda' + \Sigma_{22})^{-1} (-\Lambda\Sigma_{11} + \Sigma_{21}))(A_1^{-1})'.$$

Now the theorem is a direct consequence of $\Lambda = A_2 A_1^{-1}$. □

4.6.2 A Representation Derived from a Special Class of Admissible Matrices

The following result gives a special class of admissible matrices.

Theorem 39 *Let A be a matrix in $\mathbb{R}(m,n)$ with $rank(A) = n \leq m$ and let T_2 in $\mathbb{R}(m-n,m)$ be a matrix of rank $m-n$ such that $T_2A = 0$. Then the matrix*

$$T = \begin{pmatrix} A' \\ T_2 \end{pmatrix}$$

is admissible.

Proof. According to theorem 35, it is sufficient to show that T is regular. To do this, it is proved that the m columns of T are linearly independent, i.e. $Tu = 0$ implies $u = 0$.

But $Tu = 0$ implies that $A'u = 0$ and $T_2u = 0$ and so $u \in \mathcal{N}(T_2)$. Since the rank of T_2 is $m-n$ and $T_2A = 0$, it follows from lemma 36 that $\mathcal{N}(T_2) = \mathcal{R}(A)$. Therefore $u \in \mathcal{R}(A)$ and so there exists v such that $u = Av$. But $A'u = 0$ becomes $A'Av = 0$, which in turn implies that $v = 0$, and hence $u = 0$. □

Theorem 40 *Let T_2 in $\mathbb{R}(m-n,m)$ be a matrix of rank $m-n$ such that $T_2A = 0$. Then the density of the precise Gaussian hint inferred on x has mean*

$$m = (A'A)^{-1}A'z - (A'A)^{-1}A'\Sigma T_2'(T_2\Sigma T_2')^{-1}T_2z \qquad (4.6.18)$$

and covariance

$$V = (A'A)^{-1}(A'\Sigma A - (A'\Sigma T_2')(T_2\Sigma T_2')^{-1}T_2\Sigma A)(A'A)^{-1}.$$

Proof. Since

$$T = \begin{pmatrix} A_1' & A_2' \\ T_{21} & T_{22} \end{pmatrix} = \begin{pmatrix} A' \\ T_2 \end{pmatrix}$$

is admissible according to theorem 39, it follows that T^{-1} is a permissible basis. By theorem 33, defining

$$S = C^{-1} = (T^{-1})^{-1} = T = \begin{pmatrix} A' \\ T_2 \end{pmatrix} = \begin{pmatrix} S_1 \\ S_2 \end{pmatrix},$$

i.e.

$$S_1 = A' = \begin{pmatrix} A_1' & A_2' \end{pmatrix}$$

and

$$S_2 = T_2 = \begin{pmatrix} T_{21} & T_{22} \end{pmatrix},$$

it can be concluded that

$$m = (A'A)^{-1}A'z - (A'A)^{-1}A'\Sigma T_2'(T_2\Sigma T_2')^{-1}T_2z$$

and

$$V = (A'A)^{-1}(A'\Sigma A - (A'\Sigma T_2')(T_2\Sigma T_2')^{-1}T_2\Sigma A)(A'A)^{-1}.$$

□

It is interesting to note that the first term in the right-hand side of equation (4.6.18) is the least squares estimator of the parameter x when the covariance matrix of the error term is $\sigma^2 I$. Furthermore, from a computational point of view, the calculation of the mean m and the covariance matrix V is easier with theorem 40 than with theorem 31 because the difficult operation of inverting the $(m \times m)$ matrix Σ is replaced by two simpler operations, namely the inversion of the smaller $(n \times n)$ matrix $A'A$ and the inversion of the smaller $((m-n) \times (m-n))$ matrix $T_2 \Sigma T_2'$.

4.7 Full Rank Matrices T_2 such that $T_2 A = 0$

In this section, several methods for finding matrices T_2 in $\mathbb{R}(m-n, m)$ such that the rank of T_2 is $m-n$ and $T_2 A = 0$ are presented. These matrices can then be used to compute the mean and covariance of the inferred hint on x by using the results of theorem 40. In this section, it is still assumed that

$$A = \begin{pmatrix} A_1 \\ A_2 \end{pmatrix}$$

where $A_1 \in \mathbb{R}(n,n)$ is regular and $A_2 \in \mathbb{R}(m-n,n)$.

4.7.1 A Matrix Based on Linear Dependencies

If we define the matrix

$$\Lambda = A_2 A_1^{-1}$$

then the matrix

$$T_2 = \begin{pmatrix} -\Lambda & I_{m-n} \end{pmatrix}$$

has rank $m-n$ and $T_2 A = 0$ because

$$T_2 A = \begin{pmatrix} -A_2 A_1^{-1} & I \end{pmatrix} \begin{pmatrix} A_1 \\ A_2 \end{pmatrix} = -A_2 + A_2 = 0.$$

Furthermore, the rank of T_2 is clearly $m-n$. For example, if we take

$$A = \begin{pmatrix} 1 & 2 \\ 3 & 4 \\ 5 & 6 \end{pmatrix}, \tag{4.7.19}$$

then

$$A_1^{-1} = \begin{pmatrix} -2 & 1 \\ 1.5 & -0.5 \end{pmatrix}$$

and

$$\Lambda = A_2 A_1^{-1} = \begin{pmatrix} -1 & 2 \end{pmatrix}$$

and hence

$$T_2 = \begin{pmatrix} 1 & -2 & 1 \end{pmatrix}. \tag{4.7.20}$$

4.7.2 A Matrix Based on the Householder Method

In his book, Householder [22] has devised a method for finding an orthogonal matrix T such that

$$TA = \begin{pmatrix} U \\ 0 \end{pmatrix},$$

where U is an upper triangular matrix of order n and 0 is a matrix in $\mathbb{R}(m-n, n)$ consisting of zeros only. The method for constructing T is called the *Householder reduction method* and it is explained in Rao [36], point (vii) page 20.

Now, if we define T_2 to be the $m-n$ last rows of T, then clearly $T_2A = 0$ and the rank of T_2 is $m-n$ because T_2 is a submatrix of the regular matrix T (T is regular because $T^{-1} = T'$ since T is orthogonal).

4.7.3 A Matrix Based on Classical Variable Elimination

The classical variable elimination process to get rid of the variables $x_1, \ldots, x_n$ in the linear system $Ax = y$ allows us to find a matrix $T_2 \in \mathbb{R}(m-n, m)$ of rank $m-n$ such that $T_2A = 0$.

The determination of the matrix T_2 is illustrated by the following example where

$$A = \begin{pmatrix} 1 & 2 \\ 3 & 4 \\ 5 & 6 \end{pmatrix}$$

is the same matrix as in equation (4.7.19). Let us consider the system

$$Ax = y,$$

i.e.

$$\begin{aligned} x_1 + 2x_2 &= y_1 \\ 3x_1 + 4x_2 &= y_2 \\ 5x_1 + 6x_2 &= y_3. \end{aligned} \tag{4.7.21}$$

From the first equation of this system it is deduced that

$$x_1 = y_1 - 2x_2,$$

which can be used to rewrite the other two equations as

$$\begin{aligned} -2x_2 &= y_2 - 3y_1 \\ -4x_2 &= y_3 - 5y_1. \end{aligned} \tag{4.7.22}$$

From the first equation of this system, it follows that

$$x_2 = -0.5y_2 + 1.5y_1,$$

which can be used in the second equation to get

$$y_1 - 2y_2 + y_3 = 0. \tag{4.7.23}$$

Now, if we define the matrix

$$T_2 = \begin{pmatrix} 1 & -2 & 1 \end{pmatrix} \tag{4.7.24}$$

obtained by looking at the coefficients of the variables y_1, y_2 and y_3 in equation (4.7.23), then $T_2 A = 0$.

Note that finding a matrix T_2 by eliminating variables is equivalent, in a self-explanatory sense described below, to the selection of the last $m - n$ rows of a regular matrix T of order m such that

$$TA = \begin{pmatrix} U \\ 0 \end{pmatrix}, \tag{4.7.25}$$

where U is an upper triangular matrix of order n having all of its diagonal elements equal to 1. For such a matrix T, if we define $T_1 \in \mathbb{R}(n, m)$ and $T_2 \in \mathbb{R}(m - n, m)$ such that

$$T = \begin{pmatrix} T_1 \\ T_2 \end{pmatrix},$$

then $T_2 A = 0$ and the rank of T_2 is $m - n$ since T is regular. It is well-known from linear algebra that the matrix T is obtained by multiplication of the regular matrices corresponding to the elementary row operations needed to be applied on A in order to put it in echelon form, i.e. in the form

$$\begin{pmatrix} U \\ 0 \end{pmatrix}, \tag{4.7.26}$$

where U is a regular matrix as described above.

Let us illustrate this on the matrix A already considered above, i.e.

$$A = \begin{pmatrix} 1 & 2 \\ 3 & 4 \\ 5 & 6 \end{pmatrix}.$$

The matrix T will be found by performing the same elementary row transformations on the identity matrix I as those required to put A in the form (4.7.26). The following sequence is obtained

$$\begin{pmatrix} 1 & 2 \\ 3 & 4 \\ 5 & 6 \end{pmatrix} \qquad \begin{pmatrix} 1 & 0 & 0 \\ 0 & 1 & 0 \\ 0 & 0 & 1 \end{pmatrix} \tag{4.7.27}$$

Replace row 2 by itself minus 3 times row 1 and replace row 3 by itself minus 5 times row 1 :

$$\begin{pmatrix} 1 & 2 \\ 0 & -2 \\ 0 & -4 \end{pmatrix} \qquad \begin{pmatrix} 1 & 0 & 0 \\ -3 & 1 & 0 \\ -5 & 0 & 1 \end{pmatrix}. \tag{4.7.28}$$

Divide row 2 by -2 :

$$\begin{pmatrix} 1 & 2 \\ 0 & 1 \\ 0 & -4 \end{pmatrix} \qquad \begin{pmatrix} 1 & 0 & 0 \\ 1.5 & -0.5 & 0 \\ -5 & 0 & 1 \end{pmatrix} \tag{4.7.29}$$

Replace row 3 by itself plus 4 times row 2 :

$$\begin{pmatrix} 1 & 2 \\ 0 & 1 \\ 0 & 0 \end{pmatrix} \qquad \begin{pmatrix} 1 & 0 & 0 \\ 1.5 & -0.5 & 0 \\ 1 & -2 & 1 \end{pmatrix}. \tag{4.7.30}$$

The matrix T is the one appearing on the right-hand side of (4.7.30) and hence T_2 is the last row of this matrix, i.e.

$$T_2 = \begin{pmatrix} 1 & -2 & 1 \end{pmatrix}.$$

This is the same matrix as the one given in equations (4.7.20) and (4.7.24). With this matrix T, equation (4.7.25) holds because

$$TA = \begin{pmatrix} 1 & 2 \\ 0 & 1 \\ 0 & 0 \end{pmatrix}$$

and $T_2A = 0$.

In equations (4.7.27) to (4.7.30), let M denote the different matrices on the left-hand side, i.e. the different transformations of the matrix A obtained by elementary row transformations, and let N denote the different matrices on the right-hand side, i.e. the different transformations of the identity matrix obtained by the same elementary row transformations. Then the system (4.7.21) is nothing but

$$Mx = Ny$$

with M and N taken from (4.7.27), the system (4.7.22) is $M_1x = N_1y$ where M_1 and N_1 are the last two rows of the matrices M and N taken from (4.7.28), and equation (4.7.23) is $M_2x = N_2y$ where M_2 and N_2 are the last rows of M and N taken from (4.7.30).

4.7.4 A Matrix Based on a Generalized Inverse of A

It is well known that a matrix G in $\mathbb{R}(n, m)$ such that

$$AGA = A$$

is called a *generalized inverse* of A (see e.g. Harville [21]). The following result of linear algebra shows the structure of the family consisting of all generalized inverses of A [21].

Theorem 41 *A matrix G in $\mathbb{R}(n, m-n)$ is a generalized inverse of A if and only if there exists a matrix X in $\mathbb{R}(n, m-n)$ such that*

$$G = \begin{pmatrix} A_1^{-1} - XA_2A_1^{-1} & X \end{pmatrix}.$$

In particular, if we take $X = 0$, the matrix

$$\begin{pmatrix} A_1^{-1} & 0 \end{pmatrix}$$

is a generalized inverse of A. So the problem of finding a generalized inverse of A is trivial once we have found the inverse of A_1.

Now, if A^- denotes any generalized inverse of A, consider the matrix

$$Z = I - AA^- \in \mathbb{R}(m, m).$$

Then $ZA = 0$ because

$$ZA = IA - AA^-A = A - A = 0.$$

But by lemma 10.2.6 page 135 of Harville [21],

$$rank(Z) = m - n.$$

Therefore, there exist $(m - n)$ linearly independent rows in Z, which are assumed without loss of generality to be the first $m - n$ ones. This induces the corresponding decomposition of Z as

$$Z = \begin{pmatrix} Z_1 \\ Z_2 \end{pmatrix},$$

where $Z_1 \in \mathbb{R}(m-n, m)$ and $Z_2 \in \mathbb{R}(n, m)$. So $ZA = 0$ implies that $Z_1A = 0$ and if we take $T_2 = Z_1$ then $T_2A = 0$ and $rank(T_2) = m - n$.

As an example, let us consider again the matrix A defined above, i.e.

$$A = \begin{pmatrix} 1 & 2 \\ 3 & 4 \\ 5 & 6 \end{pmatrix}.$$

Then

$$A_1^{-1} = \begin{pmatrix} -2 & 1 \\ 1.5 & -0.5 \end{pmatrix}$$

and hence the matrix

$$A^- = \begin{pmatrix} -2 & 1 & 0 \\ 1.5 & -0.5 & 0 \end{pmatrix}$$

is a generalized inverse of A. Then

$$AA^- = \begin{pmatrix} 1 & 0 & 0 \\ 0 & 1 & 0 \\ -1 & 2 & 0 \end{pmatrix}$$

and hence

$$Z = I - AA^- = \begin{pmatrix} 0 & 0 & 0 \\ 0 & 0 & 0 \\ 1 & -2 & 1 \end{pmatrix},$$

which shows that the matrix

$$T_2 = \begin{pmatrix} 1 & -2 & 1 \end{pmatrix}$$

obtained by taking the last row of Z has rank $m - n = 1$ and $T_2A = 0$. Note that this is the same matrix as the one given in equation (4.7.20) and (4.7.24).

5. Assumption-Based Reasoning with General Gaussian Linear Systems

In this chapter, the assumption-based reasoning principle will be used to draw inferences about the parameter of a given general Gaussian linear system. These inferences will be expressed as Gaussian hints from which degrees of support and degrees of plausibility of hypotheses can be computed.

5.1 The Principle of the Inference

In this section we present the various operations needed for generating inferences about the unknown parameter of an arbitrary Gaussian linear system according to the assumption-based reasoning principle. So let

$$\mathbf{z} = \varphi(x) - \omega, \qquad \tilde{\omega} \sim N(0, \Sigma)$$

be a Gaussian linear system and let

$$p = (z, A, \Sigma), \qquad A \in \mathbb{R}(m, n)$$

denote its canonical potential. According the assumption-base reasoning principle, one has to determine the consequences of assuming that ω actually generated the observation $\mathbf{z}$. In contrast to the previous chapter, no further assumption about the Gaussian linear system is made, i.e. the potential p is no longer necessarily a potential of the third kind. In other words, p can be a potential of any of the four possible kinds introduced in definition 27. Clearly, a perturbation vector ω for which $\mathbf{z} + \omega$ is not in the range of φ cannot possibly have generated the observation $\mathbf{z}$. Therefore, the definition of an admissible perturbation given in the previous chapter applies to any kind of Gaussian linear system. Given the observation $\mathbf{z}$, the set of *admissible perturbations* is

$$\Omega = \{\omega : \mathbf{z} + \omega \in \mathcal{R}(\varphi)\} = -\mathbf{z} + \mathcal{R}(\varphi).$$

Expressed in canonical coordinates, this set is

$$\Omega = \{\omega \in \mathbb{R}^m : \exists\, x \in \mathbb{R}^n \text{ such that } \omega = -z + Ax\}.$$

Given an admissible perturbation $\omega \in \Omega$, the set of parameter values induced by z is

$$\Gamma(\omega) = \{x \in \mathcal{R}^n : z = \varphi(x) - \omega\}.$$

Note that $\Gamma(\omega)$ is not empty for all $\omega \in \Omega$. Clearly, any element in $\Gamma(\omega)$ may be the correct value of the unknown parameter under the assumption that ω is the realization of the random perturbation. But of course, the admissible perturbation that actually generated the observation z is unknown. However, before observing z, the distribution of the random perturbation is known. Since the observation z restricts the perturbations to the set Ω,this distribution has to be conditioned on Ω. Every admissible perturbation ω in Ω is an argument for the hypothesis that the correct value of the unknown parameter is in $\Gamma(\omega)$ and the conditional distribution on Ω reflects the strength of these arguments. This way of analyzing the model is a typical assumption-based reasoning process because the consequences of assuming that ω actually generated the observation z are considered.

In the following four sections, the operations described above will be performed on the four different kinds of Gaussian linear systems. Therefore, for every Gaussian linear system, there is an inferred Gaussian hint about the unknown parameter. The result of the analysis will be a Gaussian hint inferred on the canonical coordinates x of the unknown parameter x. Before proceeding to the analysis, remark that the Gaussian hint inferred from the vacuous potential $p = (0, 0, 0)$ is obviously the vacuous Gaussian hint.

5.2 The Gaussian Hint Inferred from a Gaussian Linear System of the First Kind

In this section, the Gaussian hint inferred from a Gaussian linear system of the first kind is determined. Let

$$p = (z, A, \Sigma)$$

be the canonical potential of a GLS of the first kind, i.e.

$$A \in \mathbb{R}(m, n) \qquad rank(A) = m < n.$$

Theorem 42 *The set of admissible perturbations Ω is $\mathbb{R}^m$.*

Proof. Let $\omega \in \mathbb{R}^m$. Since $rank(A) = m$, there exists m columns of A that are linearly independent. Collect them into the regular matrix B, which is assumed without loss of generality to be the first m columns of A. Defining the vector

$$x = \begin{pmatrix} B^{-1}(z + \omega) \\ 0 \end{pmatrix}$$

allows us to write

$$Ax = \begin{pmatrix} B & \cdots \end{pmatrix} \begin{pmatrix} B^{-1}(z+\omega) \\ 0 \end{pmatrix} = z + \omega.$$

Since

$$\Omega = \{\omega : \exists\ x \text{ such that } Ax = z + \omega\},$$

this shows that $\omega \in \Omega$ and so $\Omega = \mathbb{R}^n$. □

Therefore, the conditional distribution of $\tilde{\omega}$ is still $N(0, \Sigma)$ and hence the inferred Gaussian hint about x is

$$\mathcal{H} = (\mathbb{R}^m, \Sigma, \Gamma, \mathbb{R}^n) \tag{5.2.1}$$

with

$$\Gamma(\omega) = \{x \in \mathbb{R}^n : Ax = z + \omega\}.$$

The focal sets of this hint form a family of parallel linear manifolds of dimension $n - m$ covering $\mathbb{R}^n$. It can easily be verified that the Gaussian hint inferred from a potential that is equivalent to p is itself equivalent to the hint given in equation (5.2.1).

5.3 The Gaussian Hint Inferred from a Gaussian Linear System of the Second Kind

In this section, a Gaussian linear system given by the canonical potential

$$p = (z, A, \Sigma)$$

with

$$A \in \mathbb{R}(m, n) \qquad rank(A) < m < n$$

is considered. Let k denote the rank of A. The Gaussian hint on x derived from this GLS is now determined.

Since $rank(A) = k$, there are k linearly independent rows of A. By a possible rearrangement of the vectors of the canonical basis E_m, it can be assumed without loss of generality that these are the first k rows of A, which are grouped into the matrix A_1. If A_2 denotes the remaining $m - k$ rows, then

$$A = \begin{pmatrix} A_1 \\ A_2 \end{pmatrix}.$$

Since $rank(A) = k$, there exists a unique matrix Λ in $\mathbb{R}(m - k, k)$ such that

$$A_2 = \Lambda A_1. \tag{5.3.2}$$

Theorem 43 *The set of admissible perturbations Ω is a linear manifold of dimension k in $\mathbb{R}^m$ given by*

$$\Omega = \{\omega \in \mathbb{R}^m : \omega_2 - \Lambda\omega_1 = -z_2 + \Lambda z_1\} \tag{5.3.3}$$

where z_1 and ω_1 denote the first k components of z and ω respectively, and z_2 and ω_2 denote the last $m-k$ components of z and ω respectively.

Proof. Recall that

$$\Omega = \{\omega \in \mathbb{R}^m : \exists\, x \text{ such that } Ax = z + \omega\}.$$

Let $\omega \in \Omega$ and let x such that $Ax = z + \omega$. Since $Ax = z + \omega$,

$$A_1 x = z_1 + \omega_1 \quad A_2 x = z_2 + \omega_2,$$

i.e.

$$A_1 x = z_1 + \omega_1 \quad \Lambda A_1 x = z_2 + \omega_2,$$

and hence

$$\Lambda(z_1 + \omega_1) = z_2 + \omega_2,$$

i.e.

$$\omega_2 - \Lambda\omega_1 = -z_2 + \Lambda z_1,$$

which shows that ω is an element of the set on the right-hand side of (5.3.3).

Conversely, let ω such that

$$\omega_2 - \Lambda\omega_1 = -z_2 + \Lambda z_1.$$

It must be shown that there is a vector x such that $Ax = z + \omega$. Since

$$rank(A_1) = k,$$

there exist k columns of A_1 that are linearly independent. Let us assume that these are the first k ones (possibly rearrange the vectors of the basis E_n to make this true) and let us denote them by the matrix A_{11}. If A_{12} denotes the remaining $n-k$ columns of A_1, then

$$A_1 = \begin{pmatrix} A_{11} & A_{12} \end{pmatrix}$$

and A_{11} is regular. Now, if we define

$$x = \begin{pmatrix} A_1^{-1}(z_1 + \omega_1) \\ 0 \end{pmatrix},$$

then it will be proved below that

$$Ax = z + \omega.$$

Indeed,

$$Ax = z + \omega$$

if and only if

$$A_1 x = z_1 + \omega_1 \quad A_2 x = z_2 + \omega_2$$

if and only if

$$\begin{pmatrix} A_{11} & A_{12} \end{pmatrix} \begin{pmatrix} A_{11}^{-1}(z_1 + \omega_1) \\ 0 \end{pmatrix} = z_1 + \omega_1 \tag{5.3.4}$$

$$A_2 x = z_2 + \omega_2. \tag{5.3.5}$$

But (5.3.4) is satisfied and it remains to prove equation (5.3.5) :

$$A_2 x = \Lambda A_1 x = \Lambda(z_1 + \omega_1) = \Lambda z_1 + \Lambda \omega_1 = z_2 + \omega_2$$

because $\omega_2 - \Lambda\omega_1 = -z_2 + \Lambda z_1$.

It remains to show that Ω is a linear manifold of dimension k. If we define the matrix

$$U = \begin{pmatrix} -\Lambda & I_{m-k} \end{pmatrix}$$

in $\mathbb{R}(m-k, m)$, then

$$\Omega = \{\omega : U\omega = U(-z)\}.$$

Since the rank of U is $m-k$, this implies that Ω is a linear manifold of dimension $m-(m-k) = k$ in $\mathbb{R}^m$. □

Instead of expressing the GLS in canonical coordinates, which is

$$z = Ax - \omega, \quad \tilde{\omega} \sim N(0, \Sigma),$$

let us express it with respect to the basis

$$B = \begin{pmatrix} I_k & 0 \\ \Lambda & -I_{m-k} \end{pmatrix}$$

of $\mathcal{R}^m$, where Λ is the matrix defined in equation (5.3.2). Then by a well known result of linear algebra, the GLS expressed with respect to the bases B of $\mathcal{R}^m$ and E_n of $\mathcal{R}^n$ becomes

$$B^{-1}z = B^{-1}Ax - B^{-1}\omega, \quad \tilde{\omega} \sim N(0, \Sigma). \tag{5.3.6}$$

Furthermore,

$$\xi = B^{-1}\omega,$$

represents the coordinates of ω with respect to the basis B. Note that $B^{-1} = B$. Defining

$$B = B^{-1} = T = \begin{pmatrix} T_1 \\ T_2 \end{pmatrix}$$

with

$$T_1 = \begin{pmatrix} I_k & 0 \end{pmatrix} \quad T_2 = \begin{pmatrix} \Lambda & -I_{m-k} \end{pmatrix}$$

and

$$y_1 = T_1 z \quad y_2 = T_2 z \quad \xi_1 = T_1 \omega \quad \xi_2 = T_2 \omega$$

allows us to write

$$\xi = T\omega \quad T_1 A = \begin{pmatrix} A_1 \\ 0 \end{pmatrix}$$

and the system (5.3.6) becomes

$$T_1 z = A_1 x - T_1 \omega \quad T_2 z = -T_2 \omega,$$

i.e.

$$y_1 = A_1 x - \xi_1 \quad y_2 = -\xi_2. \qquad (5.3.7)$$

Expressed with respect to the bases B and E_n, the set of admissible perturbations is easily characterized by the condition $\xi_2 = -y_2$.

Theorem 44 *The set of admissible perturbations is*

$$\Omega = \{\xi \in \mathbb{R}^m : \xi_2 = -y_2\}.$$

Proof. Obviously,

$$\xi_2 = T_2 \omega = \begin{pmatrix} \Lambda & -I \end{pmatrix} \begin{pmatrix} \omega_1 \\ \omega_2 \end{pmatrix} = \Lambda \omega_1 - \omega_2$$

and

$$-y_2 = -T_2 z = -\begin{pmatrix} \Lambda & -I \end{pmatrix} \begin{pmatrix} z_1 \\ z_2 \end{pmatrix} = -(\Lambda z_1 - z_2) = z_2 - \Lambda z_1.$$

But by theorem 43

$$\Omega = \{\omega : \Lambda \omega_1 - \omega_2 = z_2 - \Lambda z_1\}$$

and so

$$\Omega = \{\xi : \xi_2 = -y_2\}.$$

□

According to this theorem and equation (5.3.7), the set of values of the parameter x induced by an admissible perturbation ξ, i.e. one for which $\xi_2 = -y_2$, is

$$\Gamma(\xi) = \{x : y_1 = A_1 x - \xi_1\}.$$

Since $\xi = T\omega$, the $N(0, \Sigma)$ distribution of $\tilde{\omega}$ is transformed into the $N(0, T\Sigma T')$ distribution of $\tilde{\xi}$, i.e $\tilde{\xi} \sim N(0, T\Sigma T')$. This distribution needs to be conditioned on the linear manifold

$$\Omega = \{\xi : \xi_2 = -y_2\}.$$

By Lauritzen [30], the distribution of $\tilde{\xi}_1$ given $\tilde{\xi}_2 = -y_2$ is

$$N_k(c_{12}c_{22}^{-1}(-y_2), c_{11} - c_{12}c_{22}^{-1}c_{21})$$

where c_{11}, c_{12}, c_{21} and c_{22} are implicitly defined by the equations

$$T\Sigma T' = \begin{pmatrix} T_1 \\ T_2 \end{pmatrix} \Sigma \left(T_1' \; T_2'\right) = \begin{pmatrix} c_{11} & c_{12} \\ c_{21} & c_{22} \end{pmatrix}.$$

If

$$\Sigma = \begin{pmatrix} \Sigma_{11} & \Sigma_{12} \\ \Sigma_{21} & \Sigma_{22} \end{pmatrix}$$

is the corresponding decomposition of Σ (i.e. for instance Σ_{11} is in $\mathbb{R}(k, k)$), then

$$\begin{aligned} c_{11} &= \Sigma_{11} \\ c_{12} &= \Sigma_{11}\Lambda' - \Sigma_{12} \\ c_{21} &= \Lambda\Sigma_{11} - \Sigma_{21} \\ c_{22} &= \Lambda\Sigma_{11}\Lambda' - \Lambda\Sigma_{12} - \Sigma_{21}\Lambda' + \Sigma_{22}. \end{aligned}$$

Theorem 45 *If rank$(A) = k < m < n$, then the inferred Gaussian hint on x is*

$$\mathcal{H} = (\mathbb{R}^k, V, \Gamma, \mathbb{R}^n)$$

where

$$V = \Sigma_{11} - ((\Sigma_{11}\Lambda' - \Sigma_{12})(\Lambda\Sigma_{11}\Lambda' - \Lambda\Sigma_{12} - \Sigma_{21}\Lambda' + \Sigma_{22})^{-1}(\Lambda\Sigma_{11} - \Sigma_{21})) \tag{5.3.8}$$

and

$$\Gamma(\omega) = \{x : A_1 x = t + \omega\}$$

with

$$t = z_1 - ((\Sigma_{11}\Lambda' - \Sigma_{12})(\Lambda\Sigma_{11}\Lambda' - \Lambda\Sigma_{12} - \Sigma_{21}\Lambda' + \Sigma_{22})^{-1}(\Lambda z_1 - z_2)) \tag{5.3.9}$$

and

$$A = \begin{pmatrix} A_1 \\ A_2 \end{pmatrix} \tag{5.3.10}$$

where $A_1 \in \mathbb{R}(k, n)$ has rank k and Λ is such that $A_2 = \Lambda A_1$.

Proof. This result follows directly from the developments given above. □

The focal sets $\Gamma(\omega)$ of this Gaussian hint are $(n-k)$-dimensional parallel linear manifolds covering $\mathbb{R}^n$. Furthermore, it can be proved that the result of the inference from any potential equivalent to p is a Gaussian hint that is equivalent to the hint given in theorem 45.

5.4 The Gaussian Hint Inferred from a Gaussian Linear System of the Third Kind

In this section, a Gaussian linear system given by the canonical potential

$$p = (z, A, \Sigma)$$

with

$$A \in \mathbb{R}(m,n) \qquad rank(A) = n \leq m$$

is considered. Two cases are distinguished in the following two theorems.

Theorem 46 *If $rank(A) = n < m$, then the inferred Gaussian hint on x is*

$$\mathcal{H} = (\mathbb{R}^n, (A'\Sigma^{-1}A)^{-1}, \Gamma, \mathbb{R}^n)$$

with

$$\Gamma(\omega) = \{x : Ix = (A'\Sigma^{-1}A)^{-1}A'\Sigma^{-1}z + \omega\} = \{(A'\Sigma^{-1}A)^{-1}A'\Sigma^{-1}z + \omega\}.$$

Proof. This is a direct consequence of theorem 31. □

Theorem 47 *If $rank(A) = n = m$, then the inferred Gaussian hint on x is*

$$\mathcal{H} = (\mathbb{R}^n, \Sigma, \Gamma, \mathbb{R}^n)$$

with

$$\Gamma(\omega) = \{x : Ax = z + \omega\}.$$

Proof. Since A is regular, the set Ω of admissible perturbations is $\mathbb{R}^n$ and the distribution $N(0, \Sigma)$ does not have to be conditioned. □

According to theorem 24 and its consequences, the density of the inferred precise Gaussian hint when

$$rank(A) = m = n$$

has mean $A^{-1}z$ and covariance $A^{-1}\Sigma(A')^{-1}$. These results can also be derived from theorem 31 because, as A is regular,

$$(A'\Sigma^{-1}A)^{-1}A'\Sigma^{-1}z = A^{-1}\Sigma(A')^{-1}A'\Sigma^{-1}z = A^{-1}z$$

and

$$(A'\Sigma^{-1}A)^{-1} = A^{-1}\Sigma(A')^{-1}.$$

Furthermore, according to section 4.4, any potential that is equivalent to p generates the same Gaussian hint on x.

5.5 The Gaussian Hint Inferred from a Gaussian Linear System of the Fourth Kind

In this section, a Gaussian linear system given by the canonical potential

$$p = (z, A, \Sigma)$$

with

$$A \in \mathbb{R}(m,n) \qquad rank(A) < n \leq m$$

is considered. Let k denote the rank of A. The Gaussian hint on x derived from this GLS is now determined.

Assume without loss of generality that the first k columns of A are linearly independent (rearrange the vectors of E_n if necessary). Let A_1 denote these first k columns and A_2 the remaining $n-k$ columns of A, so that

$$A = \begin{pmatrix} A_1 & A_2 \end{pmatrix}.$$

Since $rank(A) = k$, there exists a unique matrix Λ in $\mathbb{R}(k, m-k)$ such that

$$A_2 = A_1 \Lambda.$$

If x_1 denotes the first k components of x_1 and x_2 the remaining $n-k$ ones, then

$$z = Ax - \omega \tag{5.5.11}$$

becomes

$$\begin{pmatrix} A_1 & A_2 \end{pmatrix} \begin{pmatrix} x_1 \\ x_2 \end{pmatrix} = z + \omega,$$

i.e.

$$A_1 x_1 + A_2 x_2 = z + \omega,$$

i.e.

$$A_1 x_1 + A_1 \Lambda x_2 = z + \omega,$$

i.e.

$$A_1(x_1 + \Lambda x_2) = z + \omega.$$

Now, if we define the vector

$$y = x_1 + \Lambda x_2$$

in $\mathbb{R}^k$, then

$$A_1 y = z + \omega$$

and $\tilde{\omega} \sim N(0, \Sigma)$. Since $A_1 \in \mathbb{R}(m,k)$ and $rank(A_1) = k < m$ we are dealing with a GLS of the third kind in variables y. This implies that equation (5.5.11) is the same as

$$y = (A_1' \Sigma^{-1} A_1)^{-1} A_1' \Sigma^{-1} z + \xi_1$$

and

$$\tilde{\xi}_1 \sim N(0, (A_1' \Sigma^{-1} A_1)^{-1}).$$

Since each value of y corresponds to a whole set of values of x, namely the set

$$\{x : y = x_1 + \Lambda x_2\},$$

the set of values of x induced by the perturbation ξ_1 in $\mathbb{R}^k$ is

$$\Gamma(\xi_1) = \{x : x_1 + \Lambda x_2 = (A_1' \Sigma^{-1} A_1)^{-1} A_1' \Sigma^{-1} z + \xi_1\}.$$

If we define the matrix

$$M = \begin{pmatrix} I_k & \Lambda \end{pmatrix}$$

in $\mathbb{R}(k, n)$, then

$$\Gamma(\xi_1) = \{x : Mx = (A_1' \Sigma^{-1} A_1)^{-1} A_1' \Sigma^{-1} z + \xi_1\}.$$

Theorem 48 *If $rank(A) = k < n \leq m$, then the inferred Gaussian hint on x is*

$$\mathcal{H} = (\mathbb{R}^k, (A_1' \Sigma^{-1} A_1)^{-1}, \Gamma, \mathbb{R}^n)$$

where

$$\Gamma(\omega) = \{x : Mx = (A_1' \Sigma^{-1} A_1)^{-1} A_1' \Sigma^{-1} z + \omega\}.$$

with

$$A = \begin{pmatrix} A_1 & A_2 \end{pmatrix}, \quad A_1 \in \mathbb{R}(m, k), \quad rank(A_1) = k \tag{5.5.12}$$

and

$$M = \begin{pmatrix} I_k & \Lambda \end{pmatrix} \tag{5.5.13}$$

where $\Lambda \in \mathbb{R}(k, m-k)$ is defined by $A_2 = A_1 \Lambda$.

Proof. This result follows directly from the developments given above. □

Note that the focal sets $\Gamma(\omega)$ of this hint are parallel linear manifolds of dimension $n - k$ covering $\mathbb{R}^n$ since the rank of M is k. Furthermore, it can be verified that the Gaussian hint on x inferred from a potential equivalent to p is itself equivalent to the Gaussian hint given in theorem 48.

5.6 Canonical Proper Potentials

In the previous sections of this chapter, using the assumption-based reasoning principle, a Gaussian hint on x has been derived from a given potential. If $\mathcal{P}$ denotes the set of all potentials for x in $\mathbb{R}^n$ and $\mathcal{H}$ denotes the set of all Gaussian hints on $\mathbb{R}^n$, then this defines a mapping

$$i : \mathcal{P} \rightarrow \mathcal{H}$$

called *inference* : the Gaussian hint $i(p)$ is the result of the inference on x coming from the potential p. Of course, if p_1 is equivalent to p_2, then $i(p_1) = i(p_2)$, which means that i is a class operation. Now three theorems concerning canonical proper potentials are formulated.

Theorem 49 *Let $p = (z, A, \Sigma)$ be a canonical potential of the second kind, i.e. $rank(A) = k < m < n$. Let V, t and A_1 be the matrices defined in equations* (5.3.8), (5.3.9) *and* (5.3.10). *If we define the canonical proper potential $q = (t, A_1, V)$, then $i(p) = i(q)$.*

Proof. Since q is a potential of the first kind, the Gaussian hint inferred from q is

$$\mathcal{H} = (\mathbb{R}^k, V, \Gamma, \mathbb{R}^n)$$

with

$$\Gamma(\omega) = \{x \in \mathbb{R}^n : A_1 x = t + \omega\}.$$

But by theorem 45, the Gaussian hint inferred from p is also $\mathcal{H}$. □

Theorem 50 *Let $p = (z, A, \Sigma)$ be a canonical potential of the third kind, i.e. $rank(A) = n \leq m$. If we define the canonical proper potential*

$$q = ((A'\Sigma^{-1}A)^{-1}A'\Sigma^{-1}z, I, (A'\Sigma^{-1}A)^{-1}),$$

then $i(p) = i(q)$.

Proof. This is a direct consequence of theorems 46 and 47. □

Theorem 51 *Let $p = (z, A, \Sigma)$ be a canonical potential of the fourth kind, i.e. $rank(A) = k < n \leq m$. Let A_1 and M denote the matrices defined in equations* (5.5.12) *and* (5.5.13). *If we define the canonical proper potential*

$$q = ((A_1'\Sigma^{-1}A_1)^{-1}A_1'\Sigma^{-1}z, M, (A_1'\Sigma^{-1}A_1)^{-1}),$$

then $i(p) = i(q)$.

Proof. Since q is a potential of the first kind, the Gaussian hint inferred from q is

$$\mathcal{H} = (\mathbb{R}^k, (A_1'\Sigma^{-1}A_1)^{-1}, \Gamma, \mathbb{R}^n)$$

where

$$\Gamma(\omega) = \{x \in \mathbb{R}^n : Mx = (A_1'\Sigma^{-1}A_1)^{-1}A_1'\Sigma^{-1}z + \omega\}.$$

But by theorem 48, $\mathcal{H}$ is also the hint inferred from p. □

Note that if $p = (z, A, \Sigma)$ is a canonical potential of the first kind, i.e. $rank(A) = m < n$, then p is proper. Finally, let us mention that the inference from a proper canonical potential is very simple as shown in the next theorem.

Theorem 52 *Let* $p = (z, A, \Sigma)$ *be a canonical proper potential. Then the Gaussian hint inferred from* p *is*

$$\mathcal{H} = (\mathbb{R}^m, \Sigma, \Gamma, \mathbb{R}^n)$$

with

$$\Gamma(\omega) = \{x : Ax = z + \omega\}.$$

Proof. If $rank(A) = m < n$, then the result follows from the developments given in section 5.2. If $rank(A) = m = n$, then the result follows from theorem 47. □

6. Gaussian Hints as a Valuation System

In this chapter, it is explained how Gaussian hints can be placed in the axiomatic framework of the valuation calculus described by Shenoy and Shafer [44]. This theory provides the necessary tools and conditions for efficiently computing with so-called valuations, which are abstract generalizations of probability measures, possibility measures and belief functions. As will be shown in the next chapter, this general framework allows us to apply an efficient local propagation procedure to determine the Gaussian hint representing the knowledge about a set of variables that can be inferred from the available information. The local propagation procedure is a general technique used to simplify computations which is based on the idea that, under some conditions, it is correct to perform computations on domains that are much smaller than the original domain.

6.1 Shenoy-Shafer's Axiomatic Valuation Systems

In this section we introduce the basic elements of the axiomatic framework of Shenoy and Shafer presented in [44]. To do so, their presentation is followed quite closely. The primitive objects of the framework are a finite set of variables and a set of objects called valuations. The framework has two primitive operators, which are the marginalization and the combination. These operations act on the valuations of the system.

Let N be a finite set and let its elements be called variables. For each $h \subseteq N$ there is a set $\mathcal{V}_h$ whose elements are called *valuations* on h. Intuitively, a valuation in $\mathcal{V}_h$ is some measure of uncertainty relative to the variables in h. If $H \in \mathcal{V}_h$ is a valuation on h, then h is called the *domain* of H. Furthermore, let

$$\mathcal{V} = \cup \{\mathcal{V}_h : h \subseteq N\}$$

denote the set of all valuations. There is a mapping

$$\oplus : \mathcal{V} \times \mathcal{V} \longrightarrow \mathcal{V}$$

called *combination* such that $G \oplus H$ is a valuation on $g \cup h$ if G is a valuation on g and H is a valuation on h. Intuitively, combination corresponds to aggregation of information. More precisely, if G and H represent information about the variables in g and h respectively, then $G \oplus H$ represents the aggregated information for the variables in $g \cup h$.

We also assume that for each $h \subseteq N$ there is a mapping

$$\downarrow h : \cup \{\mathcal{V}_g : g \supseteq h\} \longrightarrow \mathcal{V}_h$$

called *marginalization to h* such that $\downarrow h(G)$, which is also denoted by $G^{\downarrow h}$, is a valuation on h when G is a valuation on g and $h \subseteq g$. The valuation $G^{\downarrow h}$ is called the *marginal of G to h*. Intuitively, marginalization corresponds to narrowing the focus of a valuation. More precisely, if G is a valuation on g representing some knowledge about the variables in g and $h \subseteq g$, then $G^{\downarrow h}$ represents the information for the variables in h that can be derived from G.

This theory of valuations takes place in the Boolean algebra consisting of all subsets of the set of variables N, but it can be generalized to any lattice as explained by Shafer in [40]. Note that for the purpose of this book there is no need to consider the notion of proper valuation defined in [44].

Now it is shown that Gaussian hints form an instance of the abstract valuation system described above. So let

$$N = \{t_1, t_2, \ldots\}$$

be a finite collection of variables such that t_i is the i-th canonical coordinate of an unknown parameter vector. For $h \subseteq N$, let n denote the number of variables contained in h, i.e. n is the cardinality of h. Then we define the set $\mathcal{V}_h$ of valuations on h to be the set of all Gaussian hints

$$\mathcal{H} = (\mathbb{R}^k, \Sigma, \Gamma, \mathbb{R}^n)$$

on $\mathbb{R}^n$ with the convention that the frame of the hint, i.e. $\mathbb{R}^n$, is the set of possible values of the variables in h. Then $\mathcal{H}$ is said to be a hint on h, or for the variables h, and h is called the *domain* of the hint $\mathcal{H}$. It remains to define the operations of marginalization and combination, which is done in the rest of this chapter.

6.2 Marginalization of Gaussian Hints

Let $x \subseteq N$ be a collection of variables and let

$$\mathcal{H} = (\mathbb{R}^k, \Sigma, \Gamma, \mathbb{R}^n)$$

with

$$\Gamma(\omega) = \{x \in \mathbb{R}^n : Ax = z + \omega\}$$

be a Gaussian hint for the variables x. Given a subset $x_1 \subseteq x$, we have to define the marginal Gaussian hint $\mathcal{H}^{\downarrow x_1}$ representing the information on x_1

that can be derived from $\mathcal{H}$. If $x_2 = x - x_1$, then the marginal Gaussian hint $\mathcal{H}^{\downarrow x_1}$ has to be defined by eliminating the variables x_2 while keeping the variables x_1. Let n_1 denote the number of variables in x_1 and n_2 the number of variables in x_2. Furthermore, after a possible rearrangement of the variables, let A_1 and A_2 denote the matrices of the partitioning of A corresponding to x_1 and x_2, i.e.

$$A = \begin{pmatrix} A_1 & A_2 \end{pmatrix}$$

where $A_1 \in \mathbb{R}(k, n_1)$ and $A_2 \in \mathbb{R}(k, n_2)$. Since, intuitively, marginalization to x_1 means focussing on x_1, in order to define the marginal hint $\mathcal{H}^{\downarrow x_1}$, the focal sets of $\mathcal{H}$ need to be projected to the variables x_1. For $\omega \in \mathbb{R}^k$, the projection of the focal set $\Gamma(\omega)$ is the set

$$\Gamma(\omega)^{\downarrow x_1} = \{x_1 \in \mathbb{R}^{n_1} : \exists\, x_2 \in \mathbb{R}^{n_2} \text{ such that } A_1 x_1 + A_2 x_2 = z + \omega\},$$

which is also denoted by $\Gamma^{\downarrow x_1}(\omega)$. In the next two subsections, the marginal hint is formally defined.

6.2.1 Marginalization of a Precise Gaussian Hint

In this case, A is a square and regular matrix in $\mathbb{R}(n, n)$ and its focal sets are

$$\Gamma(\omega) = \{x \in \mathbb{R}^n : Ax = z + \omega\}$$

for all $\omega \in \mathbb{R}^n$, i.e. every focal set is a singleton

$$\Gamma(\omega) = \{A^{-1}z + A^{-1}\omega\}.$$

Therefore, the marginalization of $\mathcal{H}$ to the variables x_1 is obviously the hint

$$\mathcal{H}^{\downarrow x_1} = (\mathbb{R}^{n_1}, B_1 \Sigma B_1', \Gamma^{\downarrow x_1}, \mathbb{R}^{n_1})$$

where B_1 is defined by

$$A^{-1} = \begin{pmatrix} B_1 \\ B_2 \end{pmatrix} \qquad B_1 \in \mathbb{R}(n_1, n)$$

and

$$\Gamma^{\downarrow x_1}(\xi) = \{x_1 \in \mathbb{R}^{n_1} : x_1 = B_1 z + \xi\}$$

for all $\xi \in \mathbb{R}^{n_1}$. This shows that $\mathcal{H}^{\downarrow x_1}$ is again a precise hint. In particular, if $A = I_n$, then $A^{-1} = I_n$ and so

$$B_1 = \begin{pmatrix} I_{n_1} & 0 \end{pmatrix}.$$

Therefore, if we define Σ_{11} to be the square matrix made of the first n_1 rows and the first n_1 columns of Σ, then

$$B_1 \Sigma B_1' = \left(I_{n_1} \; 0\right) \Sigma \begin{pmatrix} I_{n_1} \\ 0 \end{pmatrix} = \Sigma_{11}$$

and

$$\Gamma^{\downarrow x_1}(\xi) = \{z_1 + \xi\}$$

where

$$z = \begin{pmatrix} z_1 \\ z_2 \end{pmatrix}$$

with $z_1 \in \mathbb{R}^{n_1}$. In consequence, if $A = I_n$, then

$$\mathcal{H}^{\downarrow x_1} = (\mathbb{R}^{n_1}, \Sigma_{11}, \Gamma', \mathbb{R}^{n_1})$$

with

$$\Gamma'(\omega') = \{x_1 : Ix_1 = z_1 + \omega'\}.$$

6.2.2 Marginalization of a Non-Precise Gaussian Hint

Two cases are considered, namely

1. $rank(A) = k < n$ and $rank(A_2) = k$
2. $rank(A) = k < n$ and $rank(A_2) < k$.

Note that these are the only cases that need to be considered because it is impossible that both $rank(A) = k$ and $rank(A_2) > k$ are true at the same time. Let us start with the first case, i.e.

$$rank(A) = k < n \qquad rank(A_2) = k.$$

Theorem 53 *If $rank(A) = k < n$ and $rank(A_2) = k$, then*

$$\Gamma(\omega)^{\downarrow x_1} = \mathbb{R}^{n_1}$$

for all $\omega \in \mathbb{R}^k$.

Proof. Since the rank of A_2 is k, the matrix A_2 contains a sub-matrix Q such that Q^{-1} exists. This induces, without loss of generality, the following partitioning

$$A = \left(A_1 \; Q \; R\right)$$

where, by definition, Q corresponds to the variables x_2' and R to the variables x_2'' and $x_2 = x_2' \cup x_2''$. Now let $x_1 \in \mathbb{R}^{n_1}$ and it will be shown that $x_1 \in \Gamma(\omega)^{\downarrow x_1}$. Define $x_2 = (x_2', x_2'')$ by

$$x_2' = Q^{-1}(z + \omega - A_1 x_1) \quad x_2'' = 0.$$

Then, for this x_2,

$$A_1x_1 + A_2x_2 = z + \omega,$$

which shows that $x_1 \in \Gamma(\omega)^{\downarrow x_1}$ because

$$\begin{aligned} A_1x_1 + A_2x_2 &= A_1x_1 + Ax_2' + Rx_2'' = A_1x_1 + Qx_2'' \\ &= A_1x_1 + (z + \omega - A_1x_1) = z + \omega. \end{aligned}$$

□

In view of this result, the marginal hint $\mathcal{H}^{\downarrow x_1}$ is the vacuous hint for the variables x_1. Before looking at the situation where $rank(A_2) < k$, a few general results are presented because they will be useful on several occasions. Note that, in the next three theorems, we may have $m > n$.

Theorem 54 *Let $A \in \mathbb{R}(m,n)$ be a matrix such that*

$$A = \begin{pmatrix} A_1 & A_2 \end{pmatrix}$$

with $A_1 \in \mathbb{R}(m,n_1)$ and $A_2 \in \mathbb{R}(m,n_2)$. Let p denote the rank of A_2. If $p < m$, then there exists a matrix $K \in \mathbb{R}(m-p,m)$ such that $KA_2 = 0$ and $rank(K) = m - p$.

Proof. Let $\mathcal{R}(A_2)^{\perp}$ denote the orthogonal supplementary space of $\mathcal{R}(A_2)$. Since $\mathcal{R}(A_2)$ is a linear space of dimension p, the linear space $\mathcal{R}(A_2)^{\perp}$ has dimension $m - p$. By lemma 12.5.2 page 171 of Harville [21],

$$\mathcal{R}(A_2)^{\perp} = \mathcal{N}(A_2') \tag{6.2.1}$$

and so let $U \in \mathbb{R}(m, m-p)$ be a basis of $\mathcal{N}(A_2')$. Then

$$A_2'U = 0$$

and so

$$U'A_2 = 0.$$

Therefore, if we define the matrix

$$K = U'$$

in $\mathbb{R}(m-p,m)$, then $KA_2 = 0$ and $rank(K) = m - p$. □

Note that this proof provides a method for finding such a matrix K.

Theorem 55 *Let $A \in \mathbb{R}(m,n)$ be a matrix such that*

$$A = \begin{pmatrix} A_1 & A_2 \end{pmatrix}$$

with $A_1 \in \mathbb{R}(m,n_1)$ and $A_2 \in \mathbb{R}(m,n_2)$. Let p denote the rank of A_2 and suppose that $p < m$. Let K be any matrix in $\mathbb{R}(m-p,m)$ such that $KA_2 = 0$ and $rank(K) = m - p$. Then

(i) K' *is a basis of* $\mathcal{N}(A_2')$
(ii) $\mathcal{N}(K) = \mathcal{R}(A_2)$.

Proof. First (i) is proved. By equation (6.2.1), the dimension of $\mathcal{N}(A_2')$ is $m-p$. Since $A'K' = (KA_2)' = 0$ and K' is made of $m-p$ linearly independent vectors, it follows that K' is a basis of $\mathcal{N}(A_2')$.

Now (ii) is proved. The matrix K defines a linear mapping from $\mathbb{R}^m$ to $\mathbb{R}^{m-p}$ and hence

$$dim(\mathcal{N}(K)) + dim(\mathcal{R}(K)) = m,$$

which implies that

$$dim(\mathcal{N}(K) = m - dim(\mathcal{R}(K)) = m - rank(K) = m - (m-p) = p.$$

Since the columns of A_2 are all in $\mathcal{N}(K)$ and $rank(A_2) = p$, it follows that $\mathcal{N}(K) = \mathcal{R}(A_2)$. □

Theorem 56 *Let* $A \in \mathbb{R}(m,n)$ *be a matrix such that*

$$A = \begin{pmatrix} A_1 & A_2 \end{pmatrix}$$

with $A_1 \in \mathbb{R}(m,n_1)$ *and* $A_2 \in \mathbb{R}(m,n_2)$. *Let* p *denote the rank of* A_2 *and suppose that* $p < m$. *Let* K *be any matrix in* $\mathbb{R}(m-p,m)$ *such that* $KA_2 = 0$ *and* $rank(K) = m-p$ *and, for some* $z \in \mathbb{R}^m$, *let*

$$\Gamma(\omega) = \{x \in \mathbb{R}^n : Ax = z + \omega\}$$

for all $\omega \in \mathbb{R}^m$. *Then*

$$\Gamma(\omega)^{\downarrow x_1} = \{x_1 \in \mathbb{R}^{n_1} : (KA_1)x_1 = Kz + K\omega\}. \quad (6.2.2)$$

Proof. By definition,

$$\Gamma(\omega)^{\downarrow x_1} = \{x_1 \in \mathbb{R}^{n_1} : \exists\, x_2 \in \mathbb{R}^{n_2} \text{ such that } A_1x_1 + A_2x_2 = z + \omega\}. \quad (6.2.3)$$

Let x_1 be an element of the right-hand side (RHS) of equation (6.2.2), i.e.

$$KA_1x_1 = Kz + K\omega.$$

Then define the vector

$$y = z + \omega - A_1x_1.$$

This vector is in $\mathcal{N}(K)$ because

$$Ky = K(z + \omega - A_1x_1) = Kz + K\omega - (KA_1)x_1 = 0.$$

But, by theorem 55, $y \in \mathcal{R}(A_2)$ and hence there exists a vector x_2 in $\mathbb{R}^{n_2}$ such that $y = A_2x_2$. Then, for this x_2,

$$\begin{aligned} A_1x_1 + A_2x_2 &= A_1x_1 + y = A_1x_1 + z + \omega - A_1x_1 \\ &= z + \omega \end{aligned}$$

and hence x_1 is in the RHS of equation (6.2.3). Conversely, let x_1 be an element of the RHS of equation (6.2.3) and let x_2 be a vector such that

$$A_1x_1 + A_2x_2 = z + \omega.$$

Then we have

$$K(A_1x_1 + A_2x_2) = K(z + \omega)$$

and hence

$$KA_1x_1 + KA_2x_2 = Kz + K\omega$$

and hence

$$(KA_1)x_1 = Kz + K\omega,$$

which shows that x_1 is in the RHS of equation (6.2.2). □

Theorem 57 *Let $A \in \mathbb{R}(k,n)$ be a matrix such that*

$$A = \begin{pmatrix} A_1 & A_2 \end{pmatrix}$$

with $A_1 \in \mathbb{R}(k,n_1)$ and $A_2 \in \mathbb{R}(k,n_2)$ and suppose that $rank(A) = k < n$. Let K be any matrix in $\mathbb{R}(k-p,k)$ such that $KA_2 = 0$ and $rank(K) = k-p$. Then

(i) $k - p \leq n_1$,
(ii) the rank of KA_1 is $k-p$,
(iii) KA_1 is a matrix in $\mathbb{R}(k-p,n_1)$ of full row rank $k-p$.

Proof. First (i) is proved. By lemma 4.5.7 page 43 of Harville [21],

$$rank(A) \leq rank(A_1) + rank(A_2).$$

This implies that

$$k \leq rank(A_1) + p$$

and hence

$$rank(A_1) \geq k - p.$$

Since A_1 has n_1 columns, it follows that $rank(A_1) \leq n_1$. Then

$$k - p \leq rank(A_1) \leq n_1$$

and so $k - p \leq n_1$.

Now (ii) is proved. By theorem 17.5.4 page 398 of Harville [21],

$$rank(KA_1) = rank(K)$$

if and only if

$$\mathcal{R}(K') \cap \mathcal{N}(A_1') = \{0\}.$$

But theorem 55 (i) implies that $\mathcal{R}(K') = \mathcal{N}(A_2')$ and so

$$\begin{aligned}\mathcal{R}(K') \cap \mathcal{N}(A_1') &= \mathcal{N}(A_2') \cap \mathcal{N}(A_1') = \{u \in \mathbb{R}^k : A_2'u = 0, A_1'u = 0\} \\ &= \{u : A'u = 0\} = \{0\}\end{aligned}$$

because the k columns or A' are linearly independent since

$$rank(A) = rank(A') = k.$$

Furthermore,

$$rank(K) = k - p,$$

implies that $rank(KA_1) = k - p$. The assertion (iii) follows directly from (i) and (ii). □

Now let us consider a Gaussian hint

$$\mathcal{H} = (\mathbb{R}^k, \Sigma, \Gamma, \mathbb{R}^n)$$

with

$$\Gamma(\omega) = \{x : Ax = z + \omega\}$$

and suppose that

$$A = \begin{pmatrix} A_1 & A_2 \end{pmatrix}$$

with $A_1 \in \mathbb{R}(k, n_1)$, $A_2 \in \mathbb{R}(k, n_2)$ and

$$rank(A) = k < n, \qquad rank(A_2) < k.$$

In other words, the second case mentioned at the beginning of this subsection is considered. Then, in view of theorem 56, the following definition of the marginal Gaussian hint $\mathcal{H}^{\downarrow x_1}$ can be given. Let p denote the rank of A_2 and let K be any matrix in $\mathbb{R}(k-p, k)$ such that $KA_2 = 0$ and $rank(K) = k - p$. Then the marginalization of $\mathcal{H}$ to x_1 is the hint for x_1 given by

$$\mathcal{H}^{\downarrow x_1} = (\mathbb{R}^{k-p}, K\Sigma K', \Gamma^{\downarrow x_1}, \mathbb{R}^{n_1})$$

where

$$\Gamma^{\downarrow x_1}(\xi) = \{x_1 \in \mathbb{R}^{n_1} : (KA_1)x_1 = Kz + \xi\}$$

for all $\xi \in \mathbb{R}^{k-p}$. Note that $\mathcal{H}^{\downarrow x_1}$ is indeed a Gaussian hint according to theorem 57.

6.3 Vacuous Extension of Gaussian Hints

Let

$$\mathcal{H} = (\mathbb{R}^k, \Sigma, \Gamma, \mathbb{R}^n)$$

with

$$\Gamma(\omega) = \{x \in \mathbb{R}^n : Ax = z + \omega\}$$

be a Gaussian hint on $\mathbb{R}^n$ for the variables x. In order to be able to apply the local propagation procedure, we need to define a new Gaussian hint $\mathcal{H}^{\uparrow x+y}$ for the variables $x + y$ (the $+$ symbol stands for the union of two sets when their intersection is empty) representing the information conveyed by $\mathcal{H}$ with respect to the extended set of variables $x + y$. Let s denote the number of variables in y and let $m = n + s$ denote the number of variables in $x + y$.

Since $A \in \mathbb{R}(k,n)$, to define $\mathcal{H}^{\uparrow x+y}$, we need the matrix

$$B = \begin{pmatrix} A & 0 \end{pmatrix}$$

in $\mathbb{R}(k,m)$, i.e. 0 is the null matrix in $\mathbb{R}(k,s)$ corresponding to the additional variables y. Now, if we let

$$t = \begin{pmatrix} x \\ y \end{pmatrix},$$

then the *vacuous extension* of $\mathcal{H}$ to $x + y$ is defined to be the Gaussian hint

$$\mathcal{H}^{\uparrow x+y} = (\mathbb{R}^k, \Sigma, \Gamma^{\uparrow x+y}, \mathbb{R}^m)$$

where

$$\Gamma^{\uparrow x+y}(\omega) = \{t \in \mathbb{R}^m : Bt = z + \omega\}.$$

Note that this is indeed a Gaussian hint because the rank of B is still k and $k \leq m$ because $k \leq n \leq m$. The vacuously extended hint does not convey more information than $\mathcal{H}$ because some new variables have been added. The focal sets of the vacuously extended hint are made of parallel linear manifolds in $\mathbb{R}^m$ of dimension $m - k$, whereas before extension they were parallel linear manifolds of dimension $n - k$, which corresponds to an increase of $(m - k) - (n - k) = m - n = s$ dimensions, precisely the number of newly added variables.

6.4 Transport of Gaussian Hints

The marginalization and vacuous extension operations are subsumed by the more general notion of transport of a Gaussian hint. As before, let N be the set of variables considered in the valuation system and let x and y be any two non-empty subsets of N. Note that x and y need not be disjoint. Assume that x contains n variables and y contains m variables and $x \cap y$ contains k variables, so that $w = y - x$ contains $s = m - k$ variables. Of course we have

$$x \cup y = x \cup w = x + w$$

since $x \cap w = \emptyset$. Now let $\mathcal{H}$ be a Gaussian hint on $\mathbb{R}^n$ for the variables x and the goal is to define a new hint corresponding to the information induced by $\mathcal{H}$ on the variables y only. This new Gaussian hint is called the *transport* of $\mathcal{H}$ to y and is denoted by $\mathcal{H}^{|y}$. The transport of $\mathcal{H}$ to y is defined by

$$\mathcal{H}^{|y} = (\mathcal{H}^{\uparrow x+w})^{\downarrow y}.$$

Since $x + w = x \cup y$, the transport of $\mathcal{H}$ to y can also be written as

$$\mathcal{H}^{|y} = (\mathcal{H}^{\uparrow x \cup y})^{\downarrow y}.$$

With this transport operation, any Gaussian hint can be transported to any set of variables.

6.5 Projection and Composition of Potentials

In this section, we define two operations on potentials that will be needed for the definition of the combination of Gaussian hints. Let us start with the definition of the *projection* of a potential. We consider a Gaussian linear system expressed in canonical coordinates by

$$A_1 x + A_2 y = z + \omega, \qquad \tilde{\omega} \sim N(0, \Sigma)$$

where

$$A = \begin{pmatrix} A_1 & A_2 \end{pmatrix}$$

is a matrix in $\mathbb{R}(m, n)$ and $A_1 \in \mathbb{R}(m, n_1)$ and $A_2 \in \mathbb{R}(m, n_2)$. Note that it is quite possible that $m > n$. This Gaussian linear system is represented by the canonical potential $p = (z, A, \Sigma)$. In order to define the projection of this potential, the Gaussian linear system needs to be projected to the variables x. To do so, two cases are distinguished, namely

1. $rank(A_2) < m$
2. $rank(A_2) = m$.

Let us first analyze the case number 1, i.e. it is assumed that

$$rank(A_2) < m.$$

If p denotes the rank of A_2, then theorem 54 implies that there exists a matrix $K \in \mathbb{R}(m - p, m)$ such that $KA_2 = 0$ and $rank(K) = m - p$. Then, in view of theorem 56, the projection of the Gaussian linear system to x is defined as

$$KA_1 x + KA_2 y = Kz + K\omega, \qquad \tilde{\omega} \sim N(0, \Sigma),$$

i.e., defining $\xi = K\omega$,

$$KA_1 x = Kz + \xi, \qquad \tilde{\xi} \sim N(0, K\Sigma K').$$

This leads to the following definition.

Definition 35 *Let $p = (z, A, \Sigma)$ be a canonical potential with $rank(A_2) < m$. Then the projection of p to the variables x is*

$$p^{\downarrow x} = (Kz, KA_1, K\Sigma K').$$

It is important to notice that the projected potential $p^{\downarrow x}$ does not depend on the matrix K being used. In other words, if L is a matrix in $\mathbb{R}(m-p, m)$ such that $LA_2 = 0$ and $rank(L) = m - p$, then the potential $(Lz, LA_1, L\Sigma L')$ is equivalent to $(Kz, KA_1, K\Sigma K')$. Indeed, theorem 55 implies that both K' and L' are bases of $\mathcal{N}(A_2')$. Therefore, there exists a regular matrix C in $\mathbb{R}(m-p, m-p)$ such that $L' = K'C$. This implies that $L = C'K$ and hence

$$Lz = C'(Kz) \qquad LA_1 = C'(KA_1)$$
$$L\Sigma L' = C'K\Sigma K'C = C'(K\Sigma K')C,$$

which shows that the two potentials are equivalent.

Also, the projection of potentials of this kind is a class operation, i.e. if p_1 and p_2 are equivalent, then $p_1^{\downarrow x}$ and $p_2^{\downarrow x}$ are also equivalent. Indeed, let us consider two equivalent potentials $p_1 = (z, (A_1, A_2), \Sigma)$ and $p_2 = (Cz, (CA_1, CA_2), C\Sigma C')$. If K is used to compute $p_1^{\downarrow x}$, then $rank(CA_2) = p < m$ and the matrix KC^{-1} in $\mathbb{R}(m-p, m)$ can be used to compute $p_2^{\downarrow x}$ because

$$KC^{-1}(CA_2) = KA_2 = 0$$

and

$$rank(KC^{-1}) = rank(K) = m - p.$$

Therefore,

$$p_2^{\downarrow x} = (KC^{-1}Cz, KC^{-1}CA_1, KC^{-1}C\Sigma C'(C')^{-1}K')$$
$$= (Kz, KA_1, K\Sigma K') = p_1^{\downarrow x}.$$

Now let us analyze the case number 2, i.e. it is assumed that

$$rank(A_2) = m.$$

Then A_2 contains a regular sub-matrix $Q \in \mathbb{R}(m, m)$, which implies that $rank(A) = m < n$. Then, by theorem 53,

$$\Gamma(\omega)^{\downarrow x} = \mathbb{R}^{n_1}$$

for all $\omega \in \mathbb{R}^m$, which leads to the following definition.

Definition 36 *Let $p = (z, A, \Sigma)$ be a canonical potential with $rank(A_2) = m$. Then the projection of p to the variables x is the vacuous potential*

$$p^{\downarrow x} = (0, 0, 0).$$

It can easily be seen that the projection of potentials of this kind is also a class operation, which proves that projection of potentials altogether is class compatible.

If i denotes the inference mapping introduced in section 5.6, then for all potentials p it can be proved that

$$i(p)^{\downarrow x} = i(p^{\downarrow x}), \tag{6.5.4}$$

which means that inference and projection are commutative operations defined on potentials.

Now let us turn to the definition of the *composition* of several potentials for the same variables x. The composition of a collection of canonical potentials is obtained by considering the collection of associated Gaussian linear systems as a whole. Intuitively, this means that the potentials to be composed are aggregated into one single potential.

Definition 37 *Let*

$$p_i = (z_i, A_i, \Sigma_i), \qquad i = 1, \ldots, n$$

be a collection of canonical potentials for the variables x. Then the composition of these potentials is the potential

$$p = (z, A, \Sigma)$$

where

$$z = \begin{pmatrix} z_1 \\ \vdots \\ z_n \end{pmatrix} \qquad A = \begin{pmatrix} A_1 \\ \vdots \\ A_n \end{pmatrix}$$

and

$$\Sigma = \begin{pmatrix} \Sigma_1 & \ldots & 0 \\ 0 & \ldots & 0 \\ 0 & 0 & \Sigma_n \end{pmatrix}.$$

The composed potential is denoted by

$$p = \oplus_{i=1}^{n} p_i.$$

Note that this operation of composition of potentials is class compatible. Indeed, if

$$p_i' = (T_i z_i, T_i A_i, T_i \Sigma_i T_i'), \qquad i = 1, \ldots, n$$

is a collection of potentials such that p_i' is equivalent to p_i for all i, then $\oplus_{i=1}^{n} p_i$ is equivalent to $\oplus_{i=1}^{n} p_i'$ because the regular matrix

$$T = \begin{pmatrix} T_1 & \ldots & 0 \\ 0 & \ldots & 0 \\ 0 & 0 & T_n \end{pmatrix}$$

can be used to establish the equivalence. This allows us to speak of the composition of potentials without having to mention the basis of $\mathcal{R}^m$ that is being used.

Furthermore, the composition of potentials is clearly associative. Let us examine if it is also commutative. Let $p_1 = (z_1, A_1, \Sigma_1)$ and $p_2 = (z_2, A_2, \Sigma_2)$ be two canonical potentials. Then $p = p_1 \oplus p_2 = (z, A, \Sigma)$ with

$$z = \begin{pmatrix} z_1 \\ z_2 \end{pmatrix} \qquad A = \begin{pmatrix} A_1 \\ A_2 \end{pmatrix} \qquad \Sigma = \begin{pmatrix} \Sigma_1 & 0 \\ 0 & \Sigma_2 \end{pmatrix}.$$

If p is expressed with respect to a basis obtained by suitably permuting the canonical vectors, then we obtain the potential $p' = (z', A', \Sigma')$ where

$$z' = \begin{pmatrix} z_2 \\ z_1 \end{pmatrix} \qquad A' = \begin{pmatrix} A_2 \\ A_1 \end{pmatrix} \qquad \Sigma' = \begin{pmatrix} \Sigma_2 & 0 \\ 0 & \Sigma_1 \end{pmatrix}.$$

This shows that p is equivalent to p', and hence $p_1 \oplus p_2$ is equivalent to $p_2 \oplus p_1$ since $p' = p_2 \oplus p_1$. So the composition of potentials is also commutative.

6.6 Combination of Gaussian Hints Defined on the Same Domain

Let

$$\mathcal{H} = (\mathbb{R}^p, \Sigma, \Gamma, \mathbb{R}^n)$$

with

$$\Gamma(\omega) = \{x \in \mathbb{R}^n : Ax = z + \omega\}$$

be a Gaussian hint. Then, in canonical coordinates, the potential $\pi(\mathcal{H})$ is defined by

$$\pi(\mathcal{H}) = (z, A, \Sigma).$$

Of course, this defines the corresponding mapping

$$\pi : \mathcal{H} \longrightarrow \mathcal{P}.$$

Note that π is clearly a class operation. Furthermore, since $\pi(\mathcal{H})$ is actually a proper potential, theorem 52 implies that

$$i \circ \pi(\mathcal{H}) = \mathcal{H} \tag{6.6.5}$$

for all Gaussian hints $\mathcal{H}$. For $i = 1, 2$, let

$$\mathcal{H}_i = (\mathbb{R}^{k_i}, \Sigma_i, \Gamma_i, \mathbb{R}^n)$$

with

$$\Gamma_i(\omega_i) = \{x \in \mathbb{R}^n : A_i x = z_i + \omega_i\}$$

be two Gaussian hints for the variables x.

Definition 38 *The two Gaussian hints $\mathcal{H}_1$ and $\mathcal{H}_2$ are independent if the random variables $\tilde{\omega}_1$ and $\tilde{\omega}_2$ are independent.*

The goal is to define the combination of two independent Gaussian hints using the principles of Dempster's rule of combination. Accordingly, if $k = k_1 + k_2$ and $\omega = (\omega_1, \omega_2)'$ in $\mathbb{R}^k$, then the intersections

$$\Gamma(\omega) = \Gamma_1(\omega_1) \cap \Gamma_2(\omega_2)$$

need to be computed. Provided that they are not empty, these intersections are the focal sets of the combined Gaussian hint. But if we define

$$A = \begin{pmatrix} A_1 \\ A_2 \end{pmatrix} \in \mathbb{R}(k,n) \quad \text{and} \quad z = \begin{pmatrix} z_1 \\ z_2 \end{pmatrix} \in \mathbb{R}^k,$$

then

$$\Gamma(\omega) = \{x \in \mathbb{R}^n : Ax = z + \omega\}.$$

According to Dempster's rule, the prior distribution of $\tilde{\omega}$ must be conditioned on the set

$$\Omega = \{\omega \in \mathbb{R}^k : \exists\ x \text{ such that } Ax = z + \omega\}$$

to obtain the combined hint. Since $\tilde{\omega}_1$ and $\tilde{\omega}_2$ are independent, it follows that $\tilde{\omega} \sim N(0, \Sigma)$ with

$$\Sigma = \begin{pmatrix} \Sigma_1 & 0 \\ 0 & \Sigma_2 \end{pmatrix}.$$

But this operation of conditioning is exactly the one needed to compute the inferred hint from the canonical Gaussian linear system

$$Ax = z + \omega, \qquad \tilde{\omega} \sim N(0, \Sigma).$$

This means that the combined hint is the result of the inference from the canonical potential $p = (z, A, \Sigma)$. Since p is the composition of the potentials $\pi(\mathcal{H}_1)$ and $\pi(\mathcal{H}_2)$, we arrive at the following definition.

Definition 39 *Let $\mathcal{H}_1$ and $\mathcal{H}_2$ be two independent Gaussian hints. Then the combination of $\mathcal{H}_1$ and $\mathcal{H}_2$ is the hint*

$$\mathcal{H}_1 \oplus \mathcal{H}_2 = i(\pi(\mathcal{H}_1) \oplus \pi(\mathcal{H}_2)).$$

The combination of Gaussian hints is class compatible and commutative because so is the composition of potentials. As the combination of Gaussian hints essentially means forming intersections of focal sets and conditioning, the combination of Gaussian hints is also associative. In the next section, the combined Gaussian hint is explicitly computed for some special cases.

6.7 Computation of Combined Gaussian Hints

For $i = 1, 2$, let

$$\mathcal{H}_i = (\mathbb{R}^{k_i}, \Sigma_i, \Gamma_i, \mathbb{R}^n)$$

with

$$\Gamma_i(\omega_i) = \{x : A_i x = z_i + \omega_i\}$$

be a Gaussian hint for the variables x and let $k = k_1 + k_2$. If the composed potential

$$p = \pi(\mathcal{H}_1) \oplus \pi(\mathcal{H}_2) = (z, A, \Sigma)$$

has a special form or is of some special kind, then we can express the combined hint $\mathcal{H}_1 \oplus \mathcal{H}_2$ directly in terms of the potentials

$$p_i = \pi(\mathcal{H}_i) = (z_i, A_i, \Sigma_i).$$

In the following subsections, several different cases are examined.

6.7.1 Combination when the Composed Potential is of the Third Kind

First let us consider the case where $rank(A) = n = k$. Then theorem 47 implies that the combined hint is

$$\mathcal{H}_1 \oplus \mathcal{H}_2 = (\mathbb{R}^n, \Sigma, \Gamma, \mathbb{R}^n)$$

with

$$\Gamma(\omega) = \{x \in \mathbb{R}^n : Ax = z + \omega\}.$$

But we also have the following result.

Theorem 58 *If $rank(A) = n = k$, then*

$$\mathcal{H}_1 \oplus \mathcal{H}_2 = (\mathbb{R}^n, (A'\Sigma^{-1}A)^{-1}, \Gamma, \mathbb{R}^n)$$

where

$$\Gamma(\omega) = \{x \in \mathbb{R}^n : Ix = (A'\Sigma^{-1}A)^{-1}A'\Sigma^{-1}z + \omega\}.$$

Proof. This is a direct consequence of theorem 26. □

Now we are in position to formulate the following result concerning potentials of the third kind.

Theorem 59 *If the composed potential is a potential of the third kind, then*

$$\mathcal{H}_1 \oplus \mathcal{H}_2 = (\mathbb{R}^n, (A'\Sigma^{-1}A)^{-1}, \Gamma, \mathbb{R}^n)$$

where

$$\Gamma(\omega) = \{x : Ix = (A'\Sigma^{-1}A)^{-1}A'\Sigma^{-1}z + \omega\}.$$

Proof. If $rank(A) = n < k$, then the result is a consequence of theorem 46, whereas if $rank(A) = n = k$, then it is a consequence of theorem 58. □

Furthermore, the combined hint can be more explicitly expressed according to the following result.

Theorem 60 *If $rank(A) = n \leq k$, then the combined hint $\mathcal{H}_1 \oplus \mathcal{H}_2$ is precise and*

$$\mathcal{H}_1 \oplus \mathcal{H}_2 = (\mathbb{R}^n, V, \Gamma, \mathbb{R}^n)$$

where

$$\Gamma(\omega) = \{V(A_1'\Sigma_1^{-1}z_1 + A_2'\Sigma_2^{-1}z_2) + \omega\}$$

and

$$V = (A_1'\Sigma_1^{-1}A_1 + A_2'\Sigma_2^{-1}A_2)^{-1}.$$

Proof. This follows from theorem 59 because

$$\begin{aligned}(A'\Sigma^{-1}A)^{-1} &= ((A_1' \; A_2')\begin{pmatrix}\Sigma_1^{-1} & 0 \\ 0 & \Sigma_2^{-1}\end{pmatrix}\begin{pmatrix}A_1 \\ A_2\end{pmatrix})^{-1} \\ &= ((A_1'\Sigma_1^{-1} \; A_2'\Sigma_2^{-1})\begin{pmatrix}A_1 \\ A_2\end{pmatrix})^{-1} \\ &= (A_1'\Sigma_1^{-1}A_1 + A_2'\Sigma_2^{-1}A_2)^{-1}\end{aligned}$$

and

$$\begin{aligned}(A'\Sigma^{-1}A)^{-1}A'\Sigma^{-1}z &= (A_1'\Sigma_1^{-1}A_1 + A_2'\Sigma_2^{-1}A_2)^{-1}\cdot \\ &\quad (A_1' \; A_2')\begin{pmatrix}\Sigma_1^{-1} & 0 \\ 0 & \Sigma_2^{-1}\end{pmatrix}\begin{pmatrix}z_1 \\ z_2\end{pmatrix} \\ &= (A_1'\Sigma_1^{-1}A_1 + A_2'\Sigma_2^{-1}A_2)^{-1}(A_1'\Sigma_1^{-1}z_1 + A_2'\Sigma_2^{-1}z_2).\end{aligned}$$

□

6.7.2 Combination with a Precise Gaussian Hint

Suppose that $\mathcal{H}_1$ is precise, i.e.

$$\mathcal{H}_1 = (\mathbb{R}^n, \Sigma_1, \Gamma_1, \mathbb{R}^n)$$

with

$$\Gamma_1(\omega_1) = \{x : A_1x = z_1 + \omega_1\}$$

and $A_1 \in \mathbb{R}(n, n)$ is regular. In this case $rank(A) = n < k$ and hence theorem 60 can be used to compute $\mathcal{H}_1 \oplus \mathcal{H}_2$.

However, another route can be taken to find the combination of $\mathcal{H}_1$ and $\mathcal{H}_2$. It has been seen in theorem 26 that $\mathcal{H}_1$ is equivalent to the precise hint

$$\mathcal{H}_1' = (\mathbb{R}^n, A_1^{-1}\Sigma_1(A_1)^{-1'}, \Gamma_1', \mathbb{R}^n)$$

with

$$\Gamma_1'(\omega_1') = \{x \in \mathbb{R}^n : Ix = A_1^{-1}z_1 + \omega_1'\}$$

This leads to the following result.

Theorem 61 *Let $\mathcal{H}_1$ be a precise hint and let $\mathcal{H}_2$ be any hint. Then the combination of $\mathcal{H}_1$ and $\mathcal{H}_2$ is precise and*

$$\mathcal{H}_1 \oplus \mathcal{H}_2 = (\mathbb{R}^n, U, \Gamma, \mathbb{R}^n)$$

where

$$\Gamma(\omega) = \{U((A_1\Sigma_1)^{-1}z_1 + A_2'\Sigma_2^{-1}z_2) + \omega\}$$

and

$$U = (\Sigma_1^{-1} + A_2'\Sigma_2^{-1}A_2)^{-1}.$$

Proof. Since the combination of hints is class compatible, it follows that

$$\mathcal{H}_1 \oplus \mathcal{H}_2 = \mathcal{H}_1' \oplus \mathcal{H}_2.$$

The second item in the composed potential for $\mathcal{H}_1' \oplus \mathcal{H}_2$ is the matrix

$$L = \begin{pmatrix} I \\ A_2 \end{pmatrix},$$

for which $rank(L) = n < k$. The theorem then follows from theorem 60 because

$$(I'\Sigma_1^{-1}I + A_2'\Sigma_2^{-1}A_2)^{-1} = (\Sigma_1^{-1} + A_2'\Sigma_2^{-1}A_2)^{-1}$$

and

$$\begin{aligned}\Gamma(\omega) &= \{V(A_1'\Sigma_1^{-1}z_1 + A_2'\Sigma_2^{-1}z_2) + \omega\} \\ &= \{U(I_n'\Sigma_1^{-1}A_1^{-1}z_1 + A_2'\Sigma_2^{-1}z_2 + \omega\} \\ &= \{U(\Sigma_1^{-1}A_1^{-1}z_1 + A_2'\Sigma_2^{-1}z_2) + \omega\} \\ &= \{U((A_1\Sigma_1)^1 z_1 + A_2'\Sigma_2^{-1}z_2 + \omega\}.\end{aligned}$$

□

6.7.3 Combination of Two Precise Gaussian Hints

The following theorem examines the case where both hints $\mathcal{H}_1$ and $\mathcal{H}_2$ are precise.

Theorem 62 *Let $\mathcal{H}_1$ and $\mathcal{H}_2$ be two precise hints. Then*

$$\mathcal{H}_1 \oplus \mathcal{H}_2 = (\mathbb{R}^n, W, \Gamma, \mathbb{R}^n)$$

where

$$\Gamma(\omega) = \{W((A_1\Sigma_1)^{-1}z_1 + (A_2\Sigma_2)^{-1}z_2) + \omega\}$$

and

$$W = (\Sigma_1^{-1} + \Sigma_2^{-1})^{-1}.$$

Proof. According to theorem 26, the hint

$$\mathcal{H}_2' = (\mathbb{R}^n, A_2^{-1}\Sigma_2(A_2^{-1})', \Gamma_2', \mathbb{R}^n)$$

with

$$\Gamma_2'(\omega_2') = \{x : Ix = A_2^{-1}z_2 + \omega_2'\}$$

is equivalent to $\mathcal{H}_2'$. Since $\mathcal{H}_1 \oplus \mathcal{H}_2 = \mathcal{H}_1 \oplus \mathcal{H}_2'$, the result follows from theorem 61 because

$$(\Sigma_1^{-1} + I'\Sigma_2^{-1}I)^{-1} = (\Sigma_1^{-1} + \Sigma_2^{-1})^{-1}$$

and

$$\begin{aligned}\Gamma(\omega) &= \{W((A_1\Sigma_1)^{-1}z_1 + I'\Sigma_2^{-1}A_2^{-1}z_2) + \omega\} \\ &= \{W((A_1\Sigma_1)^{-1}z_1 + (A_2\Sigma_2)^{-1}z_2) + \omega\}.\end{aligned}$$

□

6.7.4 Combination when the Composed Potential is of the First Kind

If $rank(A) = k < n$, then

$$\mathcal{H}_1 \oplus \mathcal{H}_2 = (\mathbb{R}^k, \Sigma, \Gamma, \mathbb{R}^n) \tag{6.7.6}$$

where

$$\Gamma(\omega) = \{x : Ax = z + \omega\}$$

and

$$\Sigma = \begin{pmatrix} \Sigma_1 & 0 \\ 0 & \Sigma_2 \end{pmatrix}$$

because, in this case, the inference from the potential of the first kind $p = (z, A, \Sigma)$ is the Gaussian hint (6.7.6) according to section 5.2.

6.8 Combination of Hints Defined on Arbitrary Domains

For i in a finite set I, let $\mathcal{H}_i$ be a Gaussian hint for the variables h_i (the domains h_i need not be disjoint). Then we define the combination of this family of independent hints to be the Gaussian hint for the variables

$$h = \cup_{i \in I} h_i$$

given by

$$\oplus_{i \in I} \mathcal{H}_i = \oplus_{i \in I} (\mathcal{H}_i^{\uparrow h}).$$

The domain of the combined hint is the union of the domains of the hints being combined. This case of combination of hints refers to a situation that often occurs in applications because usually the available information arrives as independent pieces of uncertain knowledge about different sets of variables. This knowledge is represented by independent Gaussian hints for different sets of variables, which are then combined by Dempster's rule. The resulting hint is then marginalized to some group of variables of special interest.

7. Local Propagation of Gaussian Hints

In the previous chapter, it has been shown that Gaussian hints form a valuation system in the sense of Shenoy and Shafer. The domains of the hints are subsets of the entire finite set of variables

$$N = \{t_1, t_2, \ldots\}.$$

Let $\mathcal{J}$ be a finite collection of non empty subsets of N and suppose that for each $h \in \mathcal{J}$ there is a Gaussian hint for the variables in h. We also assume that the union of all the elements in $\mathcal{J}$ is the set N itself. Note that the pair $(N, \mathcal{J})$, or simply $\mathcal{J}$, is usually called a *hypergraph*. The hint for h is denoted by $\mathcal{H}_h$ and it is supposed to represent a piece of information coming from some Gaussian linear system. In order to get an overall representation of the available information, the individual independent pieces of information need to be pooled, which is done by combining the corresponding hints by Dempster's rule. Since the objective is to determine what the available information is saying about some group of variables h^*, the combined hint

$$\mathcal{H} = \oplus_{h \in \mathcal{J}} \mathcal{H}_h$$

is usually marginalized to some domain $h^* \in \mathcal{J}$. Of course, the marginal hint $\mathcal{H}^{\downarrow h^*}$ represents the information pertaining to the variables in h^* coming from all the available information expressed by the hints $\mathcal{H}_h, h \in \mathcal{J}$. In principle, by definition of the combination of hints, each hint $\mathcal{H}_h$ must first be vacuously extended to the domain N before the actual combination is performed. Then the combined hint can be marginalized to h^*. This solution of principle is certainly not very efficient because the combination actually takes place on the possibly large domain N. Since we are not primarily interested in the combined hint itself, but rather in its marginal to h^*, it might be possible to avoid the explicit computation of $\mathcal{H}$. This is indeed possible if the so-called axioms of Shenoy and Shafer [44] are satisfied. These axioms will allow us to find $\mathcal{H}^{\downarrow h^*}$ by means of an efficient local propagation algorithm that will be described in this chapter. But before looking into this algorithm, let us turn our attention to the axioms of Shenoy and Shafer [44].

7.1 The Axioms of Shenoy and Shafer

In this section, it will be proved that Gaussian hints satisfy the axioms of Shenoy and Shafer [44] needed for local propagation. First the combination of hints must be associative and commutative, which is true as shown in the previous chapter.

Let us start with the second local propagation axiom called the distributivity axiom. First some preparation is needed. Let x_1, x_2 and x_3 be three sets of variables that are mutually disjoint and let n_i denote the number of variables in x_i. Furthermore, let $\mathcal{H}_1$ be a Gaussian hint for the variables $x_1 + x_2$. More precisely, let

$$\mathcal{H}_1 = (\mathbb{R}^{k_1}, \Sigma_1, \Gamma_1, \mathbb{R}^{n_1+n_2})$$

with

$$\Gamma_1(\omega_1) = \{(x_1, x_2) \in \mathbb{R}^{n_1+n_2} : A_{11}x_1 + A_{12}x_2 = z_1 + \omega_1\}.$$

Also, let $\mathcal{H}_2$ be a Gaussian hint for the variables $x_2 + x_3$. More precisely, let

$$\mathcal{H}_2 = (\mathbb{R}^{k_2}, \Sigma_2, \Gamma_2, \mathbb{R}^{n_2+n_3})$$

with

$$\Gamma_2(\omega_2) = \{(x_2, x_3) \in \mathbb{R}^{n_2+n_3} : A_{21}x_2 + A_{22}x_3 = z_2 + \omega_2\}.$$

The rank of

$$A_1 = \begin{pmatrix} A_{11} & A_{12} \end{pmatrix}$$

is k_1 with $k_1 \leq n_1 + n_2$ and the rank of

$$A_2 = \begin{pmatrix} A_{21} & A_{22} \end{pmatrix}$$

is k_2 with $k_2 \leq n_2 + n_3$. The following theorem is very important because it says that Gaussian hints satisfy the second local propagation axiom of Shenoy and Shafer [44], the so-called *distributivity axiom*.

Theorem 63 *Let $\mathcal{H}_1$ and $\mathcal{H}_2$ be two Gaussian hints as described above. Then*

$$\mathcal{H}_1 \oplus \mathcal{H}_2^{\downarrow x_2} = (\mathcal{H}_1 \oplus \mathcal{H}_2)^{\downarrow x_1 \cup x_2}.$$

Proof. First suppose that $rank(A_{22}) = k_2$. If $p = (z, A, \Sigma)$ with

$$z = \begin{pmatrix} z_1 \\ z_2 \end{pmatrix} \qquad A = \begin{pmatrix} A_{11} & A_{12} & 0 \\ 0 & A_{21} & A_{22} \end{pmatrix} \qquad \Sigma = \begin{pmatrix} \Sigma_1 & 0 \\ 0 & \Sigma_2 \end{pmatrix},$$

then, by definition,

$$\mathcal{H}_1 \oplus \mathcal{H}_2 = i(p).$$

Then

$$(\mathcal{H}_1 \oplus \mathcal{H}_2)^{\downarrow x_1 \cup x_2} = i(p)^{\downarrow x_1 \cup x_2} = i(p^{\downarrow x_1 \cup x_2}).$$

Let us compute the projection of the potential p to the variables $x_1 + x_2$. By definition, if we define the matrix

$$Q = \begin{pmatrix} I_{k_1} & 0 \end{pmatrix} \in \mathbb{R}(k_1, k_1 + k_2),$$

then

$$p^{\downarrow x_1+x_2} = (Qz, QA, Q\Sigma Q') = (z_1, A_1, \Sigma_1).$$

This implies that

$$i(p^{\downarrow x_1+x_2}) = \mathcal{H}_1$$

because $p^{\downarrow x_1 \cup x_2}$ is proper, which shows that

$$(\mathcal{H}_1 \oplus \mathcal{H}_2)^{\downarrow x_1+x_2} = \mathcal{H}_1.$$

On the other hand, $\mathcal{H}_2^{\downarrow x_2}$ is the vacuous hint for x_2 because $rank(A_{22}) = k_2$, which in turn implies that $(\mathcal{H}_2^{\downarrow x_2})^{\uparrow x_1+x_2}$ is the vacuous hint for $x_1 + x_2$. So $\mathcal{H}_1 \oplus \mathcal{H}_2^{\downarrow x_2}$ is the result of the combination of $\mathcal{H}_1$ with the vacuous hint on $x_1 + x_2$. This result is of course the hint $\mathcal{H}_1$ itself, which implies that

$$\mathcal{H}_1 \oplus \mathcal{H}_2^{\downarrow x_2} = (\mathcal{H}_1 \oplus \mathcal{H}_2)^{\downarrow x_1 \cup x_2}.$$

Now suppose that $rank(A_{22}) < k_2$. Of course,

$$(\mathcal{H}_1 \oplus \mathcal{H}_2)^{\downarrow x_1+x_2} = \mathcal{H}_1.$$

We need to compute the projection of the potential p to $x_1 + x_2$. Let r denote the rank of A_{22}, so that $r < k_2$. Then, by theorem 54, there exists a matrix $T \in \mathbb{R}(k_2 - r, k_2)$ such that

$$rank(T) = k_2 - r$$

and $TA_{22} = 0$. By definition, if we define the matrix

$$Q = \begin{pmatrix} I_{k_1} & 0 \\ 0 & T \end{pmatrix} \in \mathbb{R}(k_1 + k_2 - r, k_1 + k_2),$$

then

$$p^{\downarrow x_1+x_2} = (Qz, QA_0, Q\Sigma Q')$$

with

$$A_0 = \begin{pmatrix} A_{11} & A_{12} \\ 0 & A_{21} \end{pmatrix}.$$

This implies that $p^{\downarrow x_1+x_2} = (z', A_0', \Sigma')$ where

$$z' = \begin{pmatrix} z_1 \\ Tz_2 \end{pmatrix} \qquad A_0' = \begin{pmatrix} A_{11} & A_{12} \\ 0 & TA_{21} \end{pmatrix} \qquad \Sigma' = \begin{pmatrix} \Sigma_1 & 0 \\ 0 & T\Sigma_2 T' \end{pmatrix}.$$

If t denotes this potential, then

$$(\mathcal{H}_1 \oplus \mathcal{H}_2)^{\downarrow x_1 + x_2} = i(t). \tag{7.1.1}$$

On the other hand, by definition, the hint $\mathcal{H}_2^{\downarrow x_2}$ is given by

$$\mathcal{H}_2^{\downarrow x_2} = (\mathbb{R}^{k_2 - p}, T\Sigma_2 T', \Gamma^{\downarrow x_2}, \mathbb{R}^{n_2})$$

with

$$\Gamma^{\downarrow x_2}(\xi) = \{x_2 \in \mathbb{R}^{n_2} : TA_{21}x_2 = Tz_2 + \xi\}.$$

This implies that

$$\pi((\mathcal{H}_2^{\downarrow x_2})^{\uparrow x_1 + x_2}) = (Tz_2, A'', T\Sigma_2 T')$$

with

$$A'' = \begin{pmatrix} 0 & TA_{21} \end{pmatrix}.$$

If q denotes this potential, then

$$\mathcal{H}_1 \oplus \mathcal{H}_2^{\downarrow x_2} = i(\pi(\mathcal{H}_1) \oplus \pi((\mathcal{H}_2^{\downarrow x_2})^{\uparrow x_1 + x_2})) = i(\pi(\mathcal{H}_1) \oplus q) = i(t).$$

Now, according to equation (7.1.1),

$$\mathcal{H}_1 \oplus \mathcal{H}_2^{\downarrow x_2} = (\mathcal{H}_1 \oplus \mathcal{H}_2)^{\downarrow x_1 \cup x_2}.$$

□

It is now shown that the first local propagation axiom of Shenoy and Shafer [44] is also satisfied. This axiom is called the *consonance of marginal axiom.*

Theorem 64 *Let x_1, x_2 and x_3 be three mutually disjoint sets of variables. Let*

$$\mathcal{H} = (\mathbb{R}^m, \Sigma, \Gamma, \mathbb{R}^n)$$

with

$$\Gamma(\omega) = \{x \in \mathbb{R}^n : Ax = z + \omega\}$$

be a Gaussian hint for the variables $x = x_1 \cup x_2 \cup x_3$. Then

$$(\mathcal{H}^{\downarrow x_1 \cup x_2})^{\downarrow x_1} = \mathcal{H}^{\downarrow x_1}.$$

Proof. Let p denote the potential $\pi(\mathcal{H}) = (z, A, \Sigma)$. Since $i \circ \pi = id$, it follows that $i(p) = \mathcal{H}$, which in turn implies that

$$\mathcal{H}^{\downarrow x_1} = i(p)^{\downarrow x_1} = i(p^{\downarrow x_1}).$$

On the other hand,

$$(\mathcal{H}^{\downarrow x_1 \cup x_2})^{\downarrow x_1} = (i(p)^{\downarrow x_1 \cup x_2})^{\downarrow x_1} = i(p^{\downarrow x_1 \cup x_2})^{\downarrow x_1} = i((p^{\downarrow x_1 \cup x_2})^{\downarrow x_1}).$$

So it is sufficient to show that $p^{\downarrow x_1} = (p^{\downarrow x_1 \cup x_2})^{\downarrow x_1}$, which is true because

$$\Gamma(\omega)^{\downarrow x_1} = (\Gamma(\omega)^{\downarrow x_1 \cup x_2})^{\downarrow x_1}$$

for all $\omega \in \mathbb{R}^m$. □

So the axioms of Shenoy and Shafer [44] for local propagation are indeed satisfied.

7.2 The Local Propagation Algorithm

Let $(N, \mathcal{J})$ be a hypergraph and let $\mathcal{H}_h, h \in \mathcal{J}$ be a finite collection of independent hints. In this section, we explain the local propagation algorithm of Shenoy and Shafer [44] for computing the marginal hint

$$(\oplus_{h \in \mathcal{J}} \mathcal{H}_h)^{\downarrow h^*},$$

where h^* is a particular domain in $\mathcal{J}$. This algorithm is correct because the axioms of Shenoy and Shafer are satisfied, and it is more efficient than the brute force application of the definition because the combination of hints actually takes place on smaller domains. For more information about this algorithm and its related concepts, the reader is referred to Shenoy and Shafer [44], Shafer [40], Shenoy [42, 43, 41], Kohlas & Monney [27] and Hänni [20].

Definition 40 *A tree $(\mathcal{K}, \mathcal{E})$ is called a Markov tree if the following conditions are satisfied :*

1. *$\mathcal{K}$ is a hypergraph*
2. *If $\{h, h'\} \in \mathcal{E}$, then $h \cap h' \neq \emptyset$*
3. *If h and h' are distinct nodes of $\mathcal{K}$, and t is both in h and h' then t is in every node on the unique path from h to h'.*

For the moment, let us assume that the nodes h of the hypergraph $\mathcal{J}$ can be connected in such a way that $\mathcal{M} = (\mathcal{J}, \mathcal{E})$ is a Markov tree. For example, if $N = \{t_1, t_2, \ldots, t_7\}$ and

$$\mathcal{J} = \{\{t_1, t_3\}, \{t_1, t_2\}, \{t_1, t_2, t_3, t_4\}, \{t_2, t_4, t_6, t_7\}, \{t_6\}, \{t_4, t_5\}\},$$

then the tree given in figure 7.1 is a Markov tree (to keep the notation simple, the variable t_i is identified with its index i).

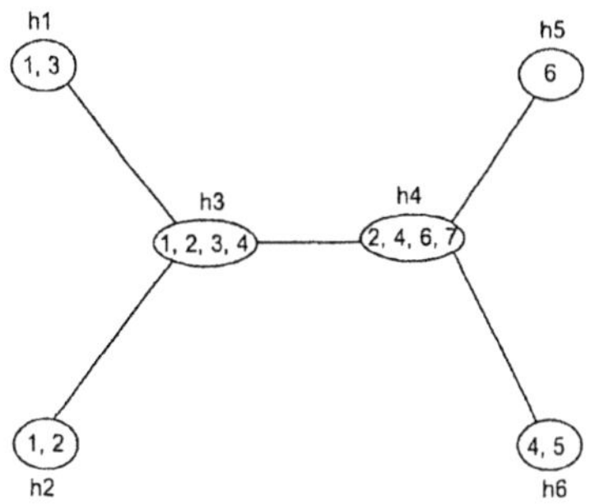

Fig. 7.1. A Markov tree

Now, instead of formally describing the algorithm by means of formulas, we simply describe it on the example introduced above, the procedure for an arbitrary Markov tree will then be obvious. Suppose that the objective is to find the marginal of the combined hint to the variables in h_3. Starting with the leaves of the tree, the hints are successively locally propagated in direction of the domain h_3 as follows :

1. Marginalize $\mathcal{H}_{h_5}$ to $h_4 \cap h_5$ and let $\mathcal{H}_1$ denote the result
2. Combine $\mathcal{H}_1$ and $\mathcal{H}_{h_4}$ (note that the combination actually takes place on the small, local domain h_4, which justifies the name "local propagation algorithm"). Let $\mathcal{H}_2$ denote the result of this combination.
3. Marginalize $\mathcal{H}_{h_6}$ to $h_4 \cap h_6$ and let $\mathcal{H}_3$ denote the result.
4. Combine $\mathcal{H}_2$ and $\mathcal{H}_3$ and let $\mathcal{H}_4$ denote the result.
5. Marginalize $\mathcal{H}_4$ to $h_4 \cap h_3$ and let $\mathcal{H}_5$ denote the result.
6. Marginalize $\mathcal{H}_{h_1}$ to $h_1 \cap h_3$ and let $\mathcal{H}_6$ denote the result.
7. Combine $\mathcal{H}_5$ and $\mathcal{H}_6$ and let $\mathcal{H}_7$ denote the result.
8. Marginalize $\mathcal{H}_{h_2}$ to $h_3 \cap h_2$ and let $\mathcal{H}_8$ denote the result.
9. Combine $\mathcal{H}_7$ and $\mathcal{H}_8$ and let $\mathcal{H}_9$ denote the result.

Then

$$(\oplus_{i=1}^{6} \mathcal{H}_{h_i})^{\downarrow h_3} = \mathcal{H}_9.$$

This procedure of local propagation in the Markov tree displayed in figure 7.1 is schematically represented in figure 7.2, where arrows symbolize marginalization and combination.

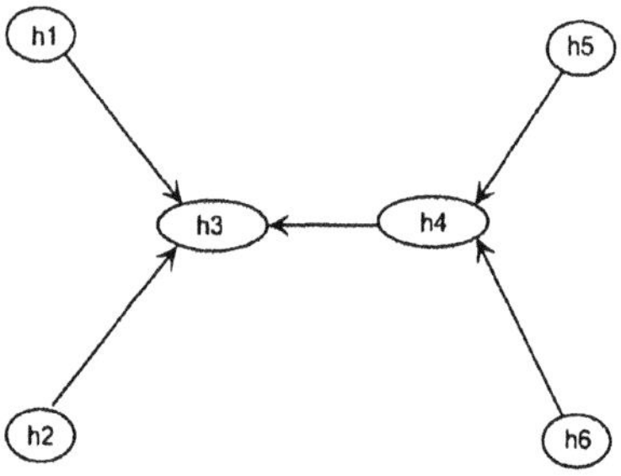

Fig. 7.2. Local propagation towards h_3

So far it has been assumed that $(\mathcal{J}, \mathcal{E})$ is a Markov tree. But it can happen that the original hypergraph $\mathcal{J}$ cannot be organized into a Markov tree by adding edges between domains in $\mathcal{J}$. In this case, some domains must be enlarged with additional variables until a Markov tree for the new domains is found. Then the usual local propagation algorithm on this so-called *covering Markov tree* must be applied. Heuristics for finding good covering Markov trees have been proposed and their optimality properties studied (see e.g. Shenoy and Shafer [44] for a list of references on this topic and Kohlas & Monney [27] for a standard method for finding covering Markov trees). With respect to local computation, let us mention in passing that Lauritzen & Jensen [45] describe an efficient propagation scheme for Bayesian networks with conditional Gaussian distributions (see also Lauritzen [29]).

In the next chapter, the local propagation algorithm described in this section will be applied to a well-known important problem in control theory.

8. Application to the Kalman Filter

One of the most important result in estimation theory in the last forty years is the Kalman filter for discrete dynamical systems [23]. Beside its theoretical appeal, this result has drawn considerable attention from practitioners and it has lead to an amazing number of applications, like, among others, marine navigation systems [3], space flight navigation, e.g. the Apollo missions to the moon [6], and telephone load forecasting [50]. The Kalman filter is typically derived from the dynamic model by using the minimum variance principle [8], or by applying Bayes rule [4], or by assuming a special form of the estimation function [5].

In this chapter, however, it will be derived from a totally different perspective, namely from the assumption-based reasoning principle that has lead to the theory of Gaussian hints as exposed in the previous chapters. This derivation is done without resorting to any minimization procedure or Bayesian procedure whatsoever, only the results from assumption-based reasoning and valuation systems will be used. A very general framework for describing and analyzing dynamical systems with stochastic disturbances or with unknown but bounded disturbances is presented in Kohlas & Monney [25].

8.1 Definition of the Random System

To start with, the dynamics of the stochastic process to be analyzed are presented. Using more or less standard notation [8], the random process to be estimated satisfies the *linear dynamic equation*

$$x(k+1) = \phi(k)x(k) + u(k), \quad k = 0, 1, 2, \ldots,$$

which is completed by the *observation process*

$$z(k) = H(k)x(k) + w(k), \quad k = 0, 1, 2, \ldots$$

where

- $x(k)$ is the *state vector* at time k, i.e. the unknown value of the random process x in $\mathbb{R}^n$ at time k
- $\phi(k)$ is a known $n \times n$-matrix relating $x(k)$ with $x(k+1)$

- $u(k)$ is a random process called *process noise.* It is assumed that

$$u(k) \sim N(0, Q(k))$$

for some known symmetric and positive definite $n \times n$-matrix $Q(k)$. In addition it is also assumed that

$$cov(u(k), u(j)) = Q(k)\delta_{kj}$$

where δ_{kj} is the Kronecker delta. These conditions mean that the random process u is *Gaussian white.* In addition, it is assumed that

$$cov(u(k), x(j)) = 0$$

for $j \leq k$
- $H(k)$ is a known $m \times n$-matrix called the *measurement matrix*
- $w(k)$ is a Gaussian white process called the *measurement noise,* i.e.

$$w(k) \sim N(0, R(k))$$

for some known $m \times m$-matrix $R(k)$, and

$$cov(w(k), w(j)) = R(k)\delta_{kj}.$$

In addition, it is supposed that

$$cov(w(k), u(j)) = 0$$

and

$$cov(w(k), x(j)) = 0$$

for all j and k.
- $z(k)$ is the known $m \times 1$-vector of observations at time k.

8.2 Derivation of the Kalman Filter

In this section, the random system described in the previous section is analyzed by means of the theory of Gaussian hints. The result of this analysis is the set of equations forming the Kalman filter. First, the measurement equation

$$z(k) = H(k)x(k) + w(k), \quad w(k) \sim N(0, R(k)) \tag{8.2.1}$$

represents a Gaussian linear system for the variables $x(k)$. In this chapter, it is assumed that all Gaussian linear systems are represented in canonical coordinates. So the canonical potential of the Gaussian linear system mentioned in (8.2.1) is

$$p_1 = (z(k), H(k), R(k)).$$

Let $\mathcal{H}(k)$ denote the hint for $x(k)$ that can be inferred from this potential, i.e. $i(p_1) = \mathcal{H}(k)$. Furthermore, the dynamics of the process expressed by the equation

$$x(k+1) = \phi(k)x(k) + u(k), \quad u(k) \sim N(0, Q(k))$$

can be rewritten as

$$-\phi(k)x(k) + x(k+1) = 0 + u(k), \quad u(k) \sim N(0, Q(k)),$$

i.e.

$$\begin{pmatrix} -\phi(k) & I \end{pmatrix} \begin{pmatrix} x(k) \\ x(k+1) \end{pmatrix} = 0 + u(k), \quad u(k) \sim N(0, Q(k)).$$

This clearly represents a Gaussian linear system for the variables

$$x(k) \cup x(k+1)$$

whose canonical potential is

$$p_2 = (0, A, Q(k))$$

with

$$A = \begin{pmatrix} -\phi(k) & I \end{pmatrix}.$$

Let $\mathcal{S}(k, k+1)$ denote the hint for

$$x(k) \cup x(k+1)$$

that can be inferred from this potential, i.e.

$$i(p_2) = \mathcal{S}(k, k+1).$$

In addition to these two different kinds of hints, we consider that some knowledge about the value of the state vector at time 0 is available, i.e knowledge about $x(0)$. In principle, this knowledge can be any Gaussian hint for the variables $x(0)$, for example the vacuous hint, but it is now assumed that it is expressed in the form of a precise hint $\mathcal{I}_0$ for the variables $x(0)$. This is a standard assumption used for initializing the Kalman filter when it is derived from the traditional minimum variance principle. If we define the potential

$$p_0 = (\mu_0, I, \Sigma_0),$$

then $i(p_0) = \mathcal{I}_0$. Figure 8.1 shows the domain of the hints $\mathcal{I}_0, \mathcal{H}(k)$ and $\mathcal{S}(k, k+1)$ that are derived from the random system. It can easily be verified that the tree obtained from this figure by dropping the diamond-shaped nodes and the arrows is indeed a Markov tree.

Now suppose that we have received the observations $z(l)$ up to time j and the goal is to infer on the value of $x(k)$ for some $k \geq j$. The result of the inference is clearly expressed by the hint

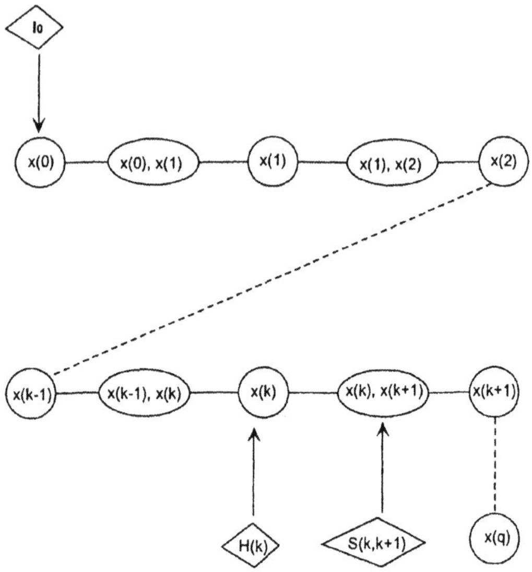

Fig. 8.1. A Markov tree with the hints in the diamond-shaped nodes

$$\hat{\mathcal{H}}(k,j) = (\mathcal{I}_0 \oplus \{\mathcal{H}(l) : l = 0, \ldots, j\} \oplus \{\mathcal{S}(s, s+1) : s = 0, \ldots, k-1\})^{|\boldsymbol{x}(k)}. \tag{8.2.2}$$

By convention, if $j = k = 0$, then

$$\oplus\{\mathcal{S}(s, s+1) : s = 0, \ldots, k-1\}$$

is defined to be the vacuous hint for $\boldsymbol{x}(0)$. If $k > j$, then a future value of the state vector is inferred as $\boldsymbol{z}(j)$ is the last observation available.

Definition 41 *If $k > j$, the hint $\hat{\mathcal{H}}(k,j)$ is called the predicted hint for $\boldsymbol{x}(k)$.*

Now suppose that we have received the observation $\boldsymbol{z}(l)$ up to time j and the goal is to infer on the value of $\boldsymbol{x}(k)$ for some time k satisfying $k \leq j$. The result of the inference is clearly expressed by the hint

$$\hat{\mathcal{H}}(k,j) = (\mathcal{I}_0 \oplus \{\mathcal{H}(l) : l = 0, \ldots, j\} \oplus \{\mathcal{S}(s, s+1) : s = 0, \ldots, j-1\})^{|\boldsymbol{x}(k)}. \tag{8.2.3}$$

If $k < j$, then a past value of the state vector is inferred as $\boldsymbol{z}(j)$ has already been observed and $k < j$.

Definition 42 *If $k < j$, the hint $\hat{\mathcal{H}}(k,j)$ is called the smoothed hint for $\boldsymbol{x}(k)$.*

Note that the definition of $\hat{\mathcal{H}}(k,j)$ depends on the relative position of k and j : it is the right-hand side of equation (8.2.2) if $k > j$ and it is the right-hand side of equation (8.2.3) if $k \leq j$. Of course, if $k = j$, then these two definitions coincide and

$$\hat{\mathcal{H}}(k,k) = (\mathcal{I}_0 \oplus \{\mathcal{H}(l) : l = 0,\ldots,k\} \oplus \{\mathcal{S}(s,s+1) : s = 0,\ldots,k-1\})^{|x(k)}.$$

Definition 43 *For all $k = 0,1,2,3,\ldots$, the hint $\hat{\mathcal{H}}(k,k)$ is called the filtered hint for $x(k)$.*

The determination of the filtered hint for $x(q)$ is particularly interesting. In other words, the objective is to compute the hint $\hat{\mathcal{H}}(q,q)$ representing the information that can be inferred on $x(q)$ from all of the available information up to time q. It turns out that this hint is precise and can be recursively computed using three equations. A series of theorems that will be used in the derivation of the Kalman filter is now formulated.

Theorem 65 *For the filtered hints we have*

$$\hat{\mathcal{H}}(0,0) = \mathcal{I}_0 \oplus \mathcal{H}(0)$$

and

$$\hat{\mathcal{H}}(k+1,k+1) = (\hat{\mathcal{H}}(k,k) \oplus \mathcal{S}(k,k+1))^{|x(k+1)} \oplus \mathcal{H}(k+1). \qquad (8.2.4)$$

Proof. The subtree consisting of the sequence of nodes between $x(0)$ and $x(k+1)$ in figure 8.1 is a Markov tree. Therefore, the local propagation algorithm can be applied to compute $\hat{\mathcal{H}}(k+1,k+1)$, which implies formula (8.2.4). On the other hand,

$$\hat{\mathcal{H}}(0,0) = (\mathcal{I}_0 \oplus \mathcal{H}(0))^{|x(0)} = \mathcal{I}_0 \oplus \mathcal{H}_0,$$

which completes the proof of the theorem. □

Theorem 66 *For all $k = 0,1,2,\ldots$, the hint $\hat{\mathcal{H}}(k,k)$ for the variables $x(k)$ and the hint $\hat{\mathcal{H}}(k,k) \oplus \mathcal{S}(k,k+1)$ for the variables $x(k) \cup x(k+1)$ are precise.*

Proof. This result is proved by induction on k. By theorem 61, first recall that the combination of a precise hint with any other hint on the same domain is again a precise hint. Since $\mathcal{I}_0$ is a precise hint for $x(0)$, theorem 65 implies that $\hat{\mathcal{H}}(0,0)$ is a precise hint for $x(0)$.

Now suppose that $\hat{\mathcal{H}}(k,k)$ is a precise hint for $x(k)$. If we define the potential

$$p_1 = \pi(\hat{\mathcal{H}}(k,k)),$$

then let

$$p_1 = (z_k, I, V_k).$$

Then the potential

$$p_1^e = (z_k, L, V_k)$$

with

$$L = (I \; 0)$$

is a potential for $x(k) \cup x(k+1)$ such that

$$p_1^e = \pi(\hat{\mathcal{H}}(k,k)^{|x(k) \cup x(k+1)}).$$

On the other hand, the potential $p_2 = (0 \;\; M \;\; Q(k))$ with

$$M = \begin{pmatrix} -\phi(k) & I \end{pmatrix}$$

for the variables $x(k) \cup x(k+1)$ is such that $p_2 = \pi(\mathcal{S}(k,k+1))$. Then, by definition,

$$\hat{\mathcal{H}}(k,k) \oplus \mathcal{S}(k,k+1) = i(p_1^e \oplus p_2)$$

and since $p_1^e \oplus p_2 = (A, z, \Sigma)$ with

$$A = \begin{pmatrix} I & 0 \\ -\phi(k) & I \end{pmatrix} \quad z = \begin{pmatrix} z_k \\ 0 \end{pmatrix} \quad \Sigma = \begin{pmatrix} V_k & 0 \\ 0 & Q(k) \end{pmatrix},$$

it follows that

$$\hat{\mathcal{H}}(k,k) \oplus \mathcal{S}(k,k+1)$$

is a precise hint for $x(k) \cup x(k+1)$ because the rank of A is $2n$ and A is a square matrix of dimension $2n$. This in turn implies that $\hat{\mathcal{H}}(k+1,k+1)$ is also precise by equation (8.2.4). □

Theorem 67 *For all $k = 1, 2, 3 \ldots$, the hint $\hat{\mathcal{H}}(k,k-1)$ is precise and*

$$\hat{\mathcal{H}}(k,k-1) = (\hat{\mathcal{H}}(k-1,k-1) \oplus \mathcal{S}(k-1,k))^{|x(k)}. \tag{8.2.5}$$

Proof. Equation (8.2.5) is a consequence of the local propagation algorithm. By theorem 66, the hint

$$\hat{\mathcal{H}}(k-1,k-1) \oplus \mathcal{S}(k-1,k)$$

is precise and hence its marginalization to $x(k)$ is also precise, thereby showing that $\hat{\mathcal{H}}(k,k-1)$ is precise. □

Theorem 68 *For all $j < k$, the hint $\hat{\mathcal{H}}(k,j)$ is precise.*

Proof. Using a similar proof as the one showing that $\hat{\mathcal{H}}(k,k) \oplus \mathcal{S}(k,k+1)$ is a precise hint for $x(k) \cup x(k+1)$ in theorem 66, it can be shown that a precise hint on $x(s)$ combined with the hint $\mathcal{S}(s,s+1)$ results in a precise hint for $x(s) \cup x(s+1)$. This result, along with the fact that the marginalization of a precise hint is still precise, implies that $\hat{\mathcal{H}}(k,j)$ is also precise when $j < k$ because $\hat{\mathcal{H}}(j,j)$ is precise by theorem 66. □

Theorem 69 *For all $j > k$, the hint $\hat{\mathcal{H}}(k,j)$ is precise.*

Proof. By the local propagation algorithm,

$$\begin{aligned}\hat{\mathcal{H}}(k,j) = \hat{\mathcal{H}}(k,k) \oplus (\oplus\{\mathcal{S}(s,s+1) : s = k,\ldots,j-1\} \\ \oplus\{\mathcal{H}(l) : l = k+1,\ldots,j\})^{|x(k)}.\end{aligned}$$

But this implies that $\hat{\mathcal{H}}(k,j)$ is precise since $\hat{\mathcal{H}}(k,k)$ is precise according to theorem 66. □

Since $\hat{\mathcal{H}}(k,j)$ is precise for all k and j, theorem 26 implies that there exists a vector $\hat{x}(k,j)$ and a matrix $P(k,j)$ such that

$$\hat{\mathcal{H}}(k,j) = (\mathbb{R}^n, P(k,j), \Gamma, \mathbb{R}^n)$$

with

$$\Gamma(\omega) = \{x(k) \in \mathbb{R}^n : Ix(k) = \hat{x}(k,j) + \omega\}$$

for all $\omega \in \mathbb{R}^n$. Then, by equation (3.3.5), the density of the precise Gaussian hint $\hat{\mathcal{H}}(k,j)$ is the normal density with mean $\hat{x}(k,j)$ and covariance $P(k,j)$, which gives a concrete interpretation to the vector $\hat{x}(k,j)$ and the matrix $P(k,j)$ because they can be used to compute degrees of support according to theorem 27.

To derive the Kalman filter, it will be necessary to consider the Gaussian linear system

$$Ix(k) = \hat{x}(k,j) + \omega, \quad \omega \sim N(0, P(k,j)).$$

Its associated canonical potential is

$$p(k,j) = (\hat{x}(k,j), I, P(k,j))$$

and by theorem 47

$$i(p(k,j)) = \hat{\mathcal{H}}(k,j).$$

In particular, for $k = j$,

$$p(k,k) = (\hat{x}(k,k), I, P(k,k))$$

and

$$Ix(k) = \hat{x}(k,k) + \omega, \quad \omega \sim N(0, P(k,k)).$$

Theorem 70 *For all* $k = 0,1,2,\ldots$, *we have*

$$P(k+1,k) = \phi(k)P(k,k)\phi(k)' + Q(k) \tag{8.2.6}$$

and

$$\hat{x}(k+1,k) = \phi(k)\hat{x}(k,k).$$

Proof. By theorem 67,

$$\hat{\mathcal{H}}(k+1,k) = (\hat{\mathcal{H}}(k,k) \oplus \mathcal{S}(k,k+1))^{|x(k+1)}.$$

Furthermore, if we define the potential

$$p_1^e = (\hat{x}(k,k), A_1, P(k,k))$$

with

$$A_1 = \begin{pmatrix} I & 0 \end{pmatrix}$$

for the variables $x(k) \cup x(k+1)$, then

$$i(p_1^e) = \hat{\mathcal{H}}(k,k)^{|x(k) \cup x(k+1)}.$$

In addition, if we define the potential $p_2 = (0 \;\; A_2 \;\; Q(k))$ with

$$A_2 = \begin{pmatrix} -\phi(k) & I \end{pmatrix}$$

for the variables $x(k) \cup x(k+1)$, then $i(p_2) = \mathcal{S}(k,k+1)$. Then, by equation (8.2.5),

$$\hat{\mathcal{H}}(k+1,k) = i(p_1^e \oplus p_2)^{|x(k+1)}.$$

But $p_1^e \oplus p_2 = (A, z, \Sigma)$ where

$$A = \begin{pmatrix} I & 0 \\ -\phi(k) & I \end{pmatrix} \quad z = \begin{pmatrix} \hat{x}(k,k) \\ 0 \end{pmatrix} \quad \Sigma = \begin{pmatrix} P(k,k) & 0 \\ 0 & Q(k) \end{pmatrix}.$$

Since A is a square matrix of dimension $2n$ and its rank is $2n$, then, by theorem 60,

$$i(p_1^e \oplus p_2) = (\mathbb{R}^{2n}, V, \Gamma, \mathbb{R}^{2n})$$

with

$$V = (A_1' P(k,k)^{-1} A_1 + A_2' Q(k)^{-1} A_2)^{-1}$$

and

$$\Gamma(\omega) = \{V(A_1' P(k,k)^{-1} \hat{x}(k,k) + A_2' Q(k)^{-1} 0 + \omega\}. \tag{8.2.7}$$

The actual computations yield

$$A_1' P(k,k)^{-1} A_1 = \begin{pmatrix} P(k,k)^{-1} & 0 \\ 0 & 0 \end{pmatrix}$$

and

$$A_2' Q(k)^{-1} A_2 = \begin{pmatrix} \phi(k)' Q(k)^{-1} \phi(k) & -\phi(k)' Q(k)^{-1} \\ -Q(k)^{-1} \phi(k) & Q(k)^{-1} \end{pmatrix}.$$

If we define A, B, C and D by the equation

$$V^{-1} = \begin{pmatrix} A & B \\ C & D \end{pmatrix},$$

then

$$A = P(k,k)^{-1} + \phi(k)'Q(k)^{-1}\phi(k)$$

$$B = -\phi(k)'Q(k)^{-1} \qquad C = -Q(k)^{-1}\phi(k) \qquad D = Q(k)^{-1}.$$

Furthermore, if we define M, N, O and P by the equation

$$V = \begin{pmatrix} M & N \\ O & P \end{pmatrix},$$

then

$$V = \begin{pmatrix} A & B \\ C & D \end{pmatrix}^{-1} = \begin{pmatrix} M & N \\ O & P \end{pmatrix}.$$

Defining $E = A - (BD^{-1}C)$ yields

$$\begin{aligned} E &= A - (-\phi(k)'Q(k)^{-1}Q(k)(-Q(k)^{-1}\phi(k))) \\ &= A - \phi(k)'Q(k)^{-1}\phi(k) \\ &= P(k,k)^{-1} + \phi(k)'Q(k)^{-1}\phi(k) - \phi(k)'Q(k)^{-1}\phi(k) \\ &= P(k,k)^{-1}. \end{aligned}$$

and defining $F = D^{-1}C$ and $G = BD^{-1}$ yield

$$F = -\phi(k) \quad G = -\phi(k)'.$$

But, by Lauritzen [30], this implies that

$$M = E^{-1} = P(k,k)$$

and

$$O = -FE^{-1} = -(-\phi(k))P(k,k) = \phi(k)P(k,k)$$

and

$$N = -E^{-1}G = -P(k,k)(-\phi(k)') = P(k,k)\phi(k)'$$

and

$$\begin{aligned} P &= D^{-1} + FE^{-1}G \\ &= Q(k) + ((-\phi(k))P(k,k)(-\phi(k)')) \\ &= Q(k) + \phi(k)P(k,k)\phi(k)'. \end{aligned}$$

So, finally,

$$V = \begin{pmatrix} P(k,k) & P(k,k)\phi(k)' \\ \phi(k)P(k,k) & Q(k) + \phi(k)P(k,k)\phi(k)' \end{pmatrix}.$$

Since by construction

$$\hat{\mathcal{H}}(k+1,k) = (\mathbb{R}^n, P(k+1,k), \Gamma', \mathbb{R}^n)$$

with

$$\Gamma'(\omega) = \{\hat{x}(k+1,k) + \omega\} \tag{8.2.8}$$

and since the marginalization of $i(p_1^e \oplus p_2)$ to $x(k+1)$ is $\hat{\mathcal{H}}(k+1,k)$, it can be concluded that

$$P(k+1,k) = \phi(k)P(k,k)\phi(k)' + Q(k),$$

which is the first part of the theorem.

Now let us turn to the second part of the theorem. By equation (8.2.7),

$$\Gamma(\omega) = \{V(A_1' P(k,k)^{-1}\hat{x}(k,k)) + \omega\}.$$

But performing the actual computations yields

$$\begin{aligned} A_1' P(k,k)^{-1}\hat{x}(k,k) &= \begin{pmatrix} I \\ 0 \end{pmatrix} P(k,k)^{-1}\hat{x}(k,k) \\ &= \begin{pmatrix} P(k,k)^{-1}\hat{x}(k,k) \\ 0 \end{pmatrix} \end{aligned}$$

and so

$$V(A_1' P(k,k)^{-1}\hat{x}(k,k)) = \begin{pmatrix} \hat{x}(k,k) \\ \phi(k)\hat{x}(k,k) \end{pmatrix}. \tag{8.2.9}$$

By marginalization to $x(k+1)$, equations (8.2.8) and (8.2.9) imply that

$$\hat{x}(k+1,k) = \phi(k)\hat{x}(k,k),$$

which completes the proof of the theorem. □

Theorem 71 *For all $k = 1, 2, 3, \ldots$, we have*

$$\hat{\mathcal{H}}(k,k) = \hat{\mathcal{H}}(k,k-1) \oplus \mathcal{H}(k)$$

Proof. This is a direct consequence of theorems 65 and 67. □

Recalling that $H(k)$ is the known measurement matrix appearing in the observation process (see the beginning of section 8.1), the following definition can be given.

Definition 44 *For $k = 1, 2, 3 \ldots$, the matrix*

$$K(k) = P(k,k-1)H(k)'(H(k)P(k,k-1)H(k)' + R(k))^{-1}$$

is called the Kalman gain matrix at time k.

Note that the Kalman filter gain matrix is a standard concept is the usual Kalman filter theory, but of course in that theory the definition of the matrix $P(k,k-1)$ (and more generally the matrix $P(k,j)$) is different from the definition of this matrix given in our theory, which is the covariance of the density of the precise Gaussian hint $\hat{\mathcal{H}}(k,k-1)$ (more generally the hint $\hat{\mathcal{H}}(k,j)$).

Theorem 72 *For all $k = 1, 2, 3\ldots$, we have*

$$P(k,k) = (I - K(k)H(k))P(k,k-1) \tag{8.2.10}$$

and

$$\hat{x}(k,k) = \hat{x}(k,k-1) - (K(k)(H(k)\hat{x}(k,k-1) - z(k))).$$

Proof. In view of theorem 71, the hint

$$\hat{\mathcal{H}}(k,k-1) \oplus \mathcal{H}(k)$$

needs to be computed. If we define the potentials

$$p_1 = (\hat{x}(k,k-1), I, P(k,k-1))$$

and

$$p_2 = (z(k), H(k), R(k)),$$

then

$$i(p_1) = \hat{\mathcal{H}}(k,k-1) \qquad i(p_2) = \mathcal{H}(k).$$

Then

$$p_1 \oplus p_2 = (z, A, \Sigma)$$

with

$$A = \begin{pmatrix} I \\ H(k) \end{pmatrix} \quad z = \begin{pmatrix} \hat{x}(k,k-1) \\ z(k) \end{pmatrix} \quad \Sigma = \begin{pmatrix} P(k,k-1) & 0 \\ 0 & R(k) \end{pmatrix}.$$

In order to compute $P(k,k)$, the hint $i(p_1 \oplus p_2)$ needs to be computed. But the rank of A is n, which is the number of columns of A, and the number of rows of A is larger than n. Then, by theorem 60,

$$P(k,k) = (P(k,k-1)^{-1} + H(k)'R(k)^{-1}H(k))^{-1}.$$

But there is also a second way of expressing $P(k,k)$. Indeed, by taking

$$A_1 = I \quad \text{and} \quad A_2 = H(k)$$

in theorem 38, we obtain

$$\begin{aligned} P(k,k) &= P(k,k-1) - (P(k,k-1)H(k)'(H(k)P(k,k-1)H(k)' \\ &\quad + R(k))^{-1}H(k)P(k,k-1)) \\ &= P(k,k-1) - (K(k)H(k)P(k,k-1)) \\ &= (I - K(k)H(k))P(k,k-1), \end{aligned}$$

which proves the first part of the theorem. Again, by theorem 38,

$$\begin{aligned}\hat{x}(k,k) &= \hat{x}(k,k-1)\\ &\quad - ((-P(k,k-1)H(k)')(H(k)P(k,k-1)H(k)' + R(k))^{-1}\\ &\quad (-H(k)\hat{x}(k,k-1) + z(k)))\\ &= \hat{x}(k,k-1) - (-K(k)(-H(k)\hat{x}(k,k-1) + z(k)))\\ &= \hat{x}(k,k-1) - (K(k)(H(k)\hat{x}(k,k-1) - z(k))),\end{aligned}$$

which completes the proof of the theorem. □

Theorem 73 *For all* $k = 0, 1, 2, 3 \ldots$, *we have*

$$\hat{x}(k+1,k+1) = \phi(k)\hat{x}(k,k) + K(k+1)(z(k+1) - H(k+1)\phi(k)\hat{x}(k,k)). \tag{8.2.11}$$

Proof. By theorem 72,

$$\hat{x}(k+1,k+1) = \hat{x}(k+1,k) - (K(k+1)(H(k+1)\hat{x}(k+1,k) - z(k+1))).$$

Then, by theorem 70,

$$\begin{aligned}\hat{x}(k+1,k+1) &= \phi(k)\hat{x}(k,k) - (K(k+1)(H(k+1)\phi(k)\hat{x}(k,k) - z(k+1)))\\ &= \phi(k)\hat{x}(k,k) + K(k+1)(z(k+1) - H(k+1)\phi(k)\hat{x}(k,k)).\end{aligned}$$

□

If we redefine $\hat{x}(k,k)$ as the best linear minimum variance estimator (BLMVE) of $x(k)$ given the observations $z(l)$ up to time k, then Kalman has shown in his seminal 1960 paper [23] that $\hat{x}(k,k)$ can be recursively computed using equations (8.2.6), (8.2.10) and (8.2.11) (this result is also proved in Catlin [8]). These equations form what is called today the *Kalman filter*.

However, the approach used in this chapter is totally different because we have combined all of the available information and marginalized it to $x(k)$. The result of the inference is the precise Gaussian hint $\hat{\mathcal{H}}(k,k)$. If the maximum density of support principle is applied to get a point estimation from the hint $\hat{\mathcal{H}}(k,k)$, i.e. the point maximizing the density of support is taken as the point estimate of $x(k)$, which is $\hat{x}(k,k)$, then it is striking that we get the same answer as the one obtained by Kalman through the minimum variance principle.

If the process and measurement noise covariances are known in advance, it is interesting to note that the sequence of Kalman gain matrices can be calculated before any measurement is taken. Then equation (8.2.11) can be used to compute the sequence of estimations $\hat{x}(k,k)$ as the observations become available.

Finally, I would like to mention that the idea of local computations for the Kalman filter has already been suggested by Dempster in the articles [16, 15]. In this chapter, using the theory of Gaussian hints developed in this book, the equations forming the Kalman filter have been explicitly derived.

References

1. R.H. Bartlett, D.W. Roloff, R.G. Cornell, A.F. Andrews, P.W. Dillon, and J.B. Zwischenberger, *Extracorporeal Circulation in Neonatal Respiratory Failure : A Prospective Randomized Study*, Pediatrics **76** (1948), 479–487.
2. C.R. Begg, *On Inferences from Wei's Biased Coin Design for Clinical Trials (with discussion)*, Biometrika **77** (1990), 467–484.
3. B.E. Bona and R.J. Smay, *Optimum Reset of Ships's Inertial Navigation System*, IEEE Transactions on Aerospace and Electronic Systems **AES-2** (1966), no. 4, 409–414.
4. L.D. Broemeling, *Bayesian Analysis of Linear Models*, Marcel Dekker, New York, 1985.
5. R.G. Brown, *Introduction to Random Signal Analysis and Kalman Filtering*, Wiley, New York, 1983.
6. R.S. Bucy and P.D. Joseph, *Filtering for Stochastic Processes with Applications to Guidance*, Interscience, New York, 1968.
7. H. Bunke, *Statistical Inference: Fiducial and Structural vs. Likelihood.*, Math. Operationsforsch. u. Statist. **6** (1975), 667–676.
8. D.E. Catlin, *Estimation, Control, and the Discrete Kalman Filter*, Applied Mathematical Sciences, vol. 71, Springer, 1989.
9. R.G. Cowell, A.P. Dawid, S.L. Lauritzen, and D.J. Spiegelhalter, *Probabilistic Networks and Expert Systems*, Statistics for Engineering and Information Science, Springer, 1999.
10. A.P. Dawid and M. Stone, *The Functional-Model Basis of Fiducial Inference*, The Annals of Statistics **10** (1982), no. 4, 1054–1067.
11. A.P. Dempster, *New Methods for Reasoning Towards Posterior Distributions Based on Sample Data.*, Annals of Math. Statistics **37** (1965), 355–374.
12. A.P. Dempster, *Upper and Lower Probability Induced by a Multivalued Mapping*, Annals of Math. Stat. **38** (1967), 325–339.
13. A.P. Dempster, *Upper and Lower Probability Inferences Based on a Sample from a Finite Univariate Population*, Biometrika **54** (1967), 515–528.
14. A.P. Dempster, *A Generalization of Bayesian Inference*, J. R. Statistical Society **30** (1968), 205–247.
15. A.P. Dempster, *Construction and Local Computation Aspects of Network Belief Functions*, Influence Diagrams, Belief Nets and Decision Analysis (Smith J.Q. Oliver R.M., ed.), 1990, pp. 121–141.
16. A.P. Dempster, *Normal Belief Functions and the Kalman Filter*, Tech. report, Department of Statistics, Harvard University, 1990.
17. R.A. Fisher, *Inverse Probability*, Proceedings Cambridge Philosophical Society **26** (1930), 528–535.
18. D.A. Fraser, *The Structure of Inference*, Wiley, 1968.
19. I. Hacking, *Logic of Statistical Inference*, Cambridge University Press, 1965.

20. R. Hänni, *Probabilistic Argumentation Systems and Symbolic Evidence Theory*, Ph.D. thesis, Institute of Informatics of the University of Fribourg (Switzerland), 1996.
21. D.A. Harville, *Matrix Algebra From a Statistician's Perspective*, Springer, 1997.
22. A.S. Householder, *The Theory of Matrices in Numerical Analysis*, Blaisdell, 1964.
23. R.E. Kalman, *A New Approach to Linear Filtering and Prediction*, Journal of Basic Engineering (ASME) **82D** (1960), 35–45.
24. F. Klawonn and P. Smets, *The Dynamic of Belief in the Transferable Belief Model and Specialization-Generalization Matrices*, Proceedings of the 8th Conference on Uncertainty in Artificial Intelligence (Dubois D., Wellman M., D'Ambrosio B., and Smets P., eds.), Morgan Kaufman Publishers, San Francisco, California, 1992, pp. 130–137.
25. J. Kohlas, *Describing Uncertainty in Dynamical Systems by Uncertain Restrictions*, Modeling, Estimation and Control of Systems with Uncertainty. (G.B. Di Masi et al., ed.), Birkhäuser, 1991, pp. 210–223.
26. J. Kohlas and P.A. Monney, *Representation of Evidence by Hints*, Advances in the Dempster-Shafer Theory of Evidence (R.R. Yager, J. Kacprzyk, and M. Fedrizzi, eds.), Wiley, 1994, pp. 473–492.
27. J. Kohlas and P.A. Monney, *A Mathematical Theory of Hints. An Approach to the Dempster-Shafer Theory of Evidence*, Lecture Notes in Economics and Mathematical Systems, vol. 425, Springer Verlag, 1995.
28. K.B. Laskey and P.E. Lehner, *Assumptions, Belief and Probabilities*, Artif. Intell. **41** (1989), 65–77.
29. S.L. Lauritzen, *Propagation of Probabilities, Means and Variances in Mixed Graphical Association Models*, J. of the American Statistical Association **87** (1992), 1098–1108.
30. S.L. Lauritzen, *Graphical models*, Oxford Statistical Science Series 17, Oxford Science Publications, 1996.
31. L. Liu, *A Theory of Gaussian Belief Functions*, Int. J. of Approximate Reasoning **14** (1996), 95–126.
32. P.A. Monney, *Dempster Specialization Matrices and the Combination of Belief Functions*, Proceedings of the Sixth European Conference on Symbolic and Quantitative Approaches to Reasoning with Uncertainty (ECSQARU), 2001, Toulouse, France, Lecture Notes in Computer Science, Springer Verlag, September 2001, pp. 316–327.
33. J. Pearl, *Probabilistic Reasoning in Intelligent Systems: Networks of Plausible Inference.*, second printing ed., Morgan Kaufmann Publ. Inc., 1991.
34. J. Pearl, *Causation, Action and Counterfactuals*, Logic and Scientific Methods (M.L. Dalla Chiara et al., ed.), Kluwer Academic Publishers, The Netherlands, 1997, pp. 355–375.
35. A. Plante, *On the Validation of Fiducial Techniques*, Canadian Journal of Statistics **7** (1979), 217–226.
36. C.R. Rao, *Linear Statistical Inference and its Applications*, Wiley, 1973.
37. R. Royall, *Statistical Evidence. A Likelihood Paradigm*, Monographs on Statistics and Applied Probability, vol. 71, Chapman & Hall, 1997.
38. G. Shafer, *A Mathematical Theory of Evidence*, Princeton University Press, 1976.
39. G. Shafer, *Belief Functions and Parametric Models*, J.R. Statistical Society B **44** (1982), no. 3, 322–352.
40. G. Shafer, *An Axiomatic Study of Computation in Hypertrees*, Working Paper 232, University of Kansas School of Business, 1991.

41. P. Shenoy, *A Valuation-Based Language for Expert Systems*, International Journal of Approximate Reasoning **3** (1989), 383–411.
42. P. Shenoy, *Independance in Valuation-Based Systems*, Working Paper 236, University of Kansas School of Business, 1991.
43. P. Shenoy, *Valuation-Based Systems: A Framework for Managing Uncertainty in Expert Systems*, Working Paper 226, University of Kansas School of Business, 1991.
44. P. Shenoy and G. Shafer, *Axioms for Probability and Belief-Function Propagation*, Uncertainty in Artificial Intelligence (Shachter R.D., Levitt T.S., Kanal L.N., and Lemmer J.F., eds.), Elsevier Science, 1990, pp. 169–198.
45. F. Jensen S.L. Lauritzen, *Stable Local Computation with Conditional Gaussian Distributions*, Tech. Report R-99-2014, Dept. of Math. Sciences, Aalborg University, 1999.
46. P. Smets, *Un Modèle Mathématico-Statistique Simulant le Processus du Diagnostic Médical*, Ph.D. thesis, Free University of Brussels, 1978.
47. P. Smets, *The Combination of Evidence in the Transferable Belief Model*, IEEE Trans. PAMI **12** (1990), 447–458.
48. P. Smets, *The Nature of the Unnormalized Beliefs Encountered in the Transferable Belief Model*, Proceedings of the 8th Conference on Uncertainty in Artificial Intelligence (Dubois D., Wellman M., D'Ambrosio B., and Smets P., eds.), Morgan Kaufman Publishers, San Francisco, California, 1992, pp. 292–297.
49. P. Smets, *Belief Induced by the Partial Knowledge of the Probabilities*, Uncertainty in Artificial Intelligence (Mantaras R.L. and Poole D., eds.), Morgan Kaufman Publisher, 1994, pp. 523–530.
50. C.R. Szelag, *A Short-Term Forecasting Algorithm for Truck Demand Servicing*, The Bell System Technical Journal **61** (1982), no. 1, 67–96.
51. J.H. Ware, *Investigating Therapies of Potentially Great Benefit: ECMO (with discussion)*, Statistical Science 4 (1989), 298–340.

Index

To Brenda

To Brenda

To Brenda

To Brenda

GPSR Compliance

The European Union's (EU) General Product Safety Regulation (GPSR) is a set of rules that requires consumer products to be safe and our obligations to ensure this.

If you have any concerns about our products, you can contact us on ProductSafety@springernature.com

In case Publisher is established outside the EU, the EU authorized representative is:

Springer Nature Customer Service Center GmbH
Europaplatz 3
69115 Heidelberg, Germany

Batch number: 10725981

Printed by Printforce, the Netherlands